Tomorrow's World

Second Volume

Tomorrow's

Second Volume

Raymond Baxter James Burke

Research by Alan Dobson/John Harrison/Louise Panton
John Parry/Robert Symes Schutzmann
William Woollard, Tomorrow's World
Reporter

World

Edited by
Michael Latham

British Broadcasting Corporation

The authors wish to record their appreciation of the research workers, scientists, engineers, and doctors who helped in the preparation of this book. Together with universities, hospitals, research establishments, and organisations all over the world, they generously offered private research material, illustrations and advice without which this book, or indeed the television programme whose name it bears, could not exist.

Title page photograph shows
the tongue of a sea slug (see page 139).

Published by the
British Broadcasting Corporation
35 Marylebone High Street
London W1M 4AA

ISBN 0 563 12017 7

First published 1971
© Raymond Baxter, James Burke,
Michael Latham 1971

Printed in England by
Jolly and Barber Limited,
Rugby, Warwickshire

To Sylvia and Madeline – and all those who put up with us while we wrote it.

Contents

Nearly everyone can remember when a human heart transplant was the stuff of lurid fiction; when a visit to the wastelands of the Moon was wild fantasy; when ideas like plastic eyes that restore sight to the blind, miniature electronic brains, or man-made diamonds and rubies were dreams for schoolboys and mad scientists. Yet they all came true. What lies ahead now?

For each one of us, tomorrow's world begins only a few brief moments away. And of one thing we can be sure. With every passing month, year and decade it will continue to reveal its quota of astonishing developments which ensures that life on Planet Earth will never be quite the same again.

It changed beyond recall when men were able to gaze upon Earth itself, glittering like a giant blue sapphire set in the velvet infinity of deep space. For, along with the incredible beauty, there were the tell-tale signs of real danger – man-made pollution of the oceans, the land masses, of the atmosphere itself. Dramatic information like that is bound to affect the way we live from now on. It was just one instance of how the impact of science and technology may colour attitudes and alter life styles.

But there are yet more fundamental changes for mankind on the horizon as you will surely discover – if you are prepared to explore the secrets of all our futures in tomorrow's world.

Skylab, the orbiting laboratory that will circle the Earth. From here scientists will observe the solar system, survey planet Earth, and study the physiological effects of space on their own bodies.

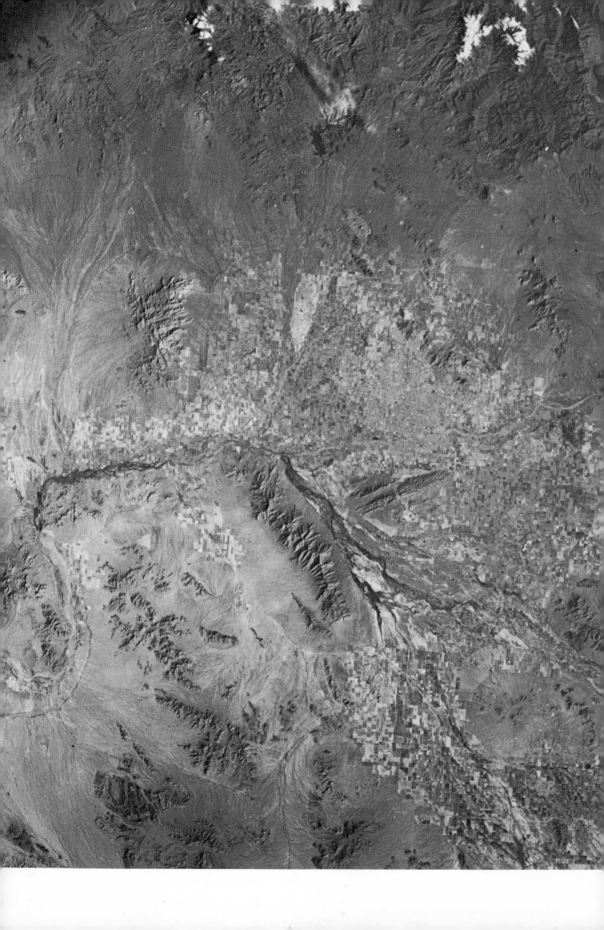

Left: Evidence of Man's presence. This infrared colour photograph of an area near Pheonix, Arizona, was taken to provide information on earth resources. The tiny red squares are fields under cultivation (the plants give off heat). Note the clustering of urban and agricultural areas along the rivers.

Top right: High-altitude weather. Looking southeast is the United Arab Republic, Arabia, the Red Sea and the Nile Valley. The strip of cloud reveals the presence of a high-altitude jet stream of warmer air. These air flows travel at speeds over 120 mph and have broken up aircraft which have flown into them.

Bottom right: Pollution or silt? This orbital photograph shows the coast off South Carolina at Cape Hatteras. The bright line in the centre is a beach thrown up some way from the mainland by wave action. The lighter blue area to seaward of the beaches is the shallow waters of the American continental shelf. Superimposed on them, the very light blue 'plumes' of discharge from a river (off picture to the left). Whether they are silt or pollutants suspended in the water is not yet known.

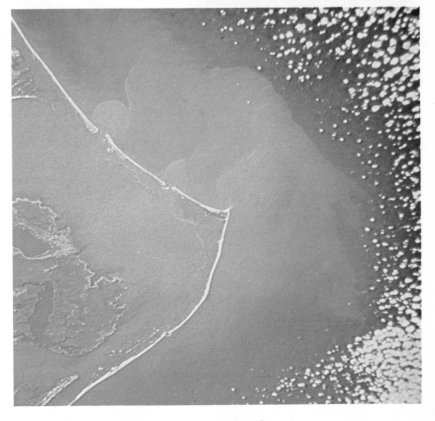

1

The Wild Black Yonder

If you take the lift to the third floor of the Mission Control building at the Manned Spacecraft Centre in Houston, Texas, you step out into a grey rubber-floored corridor. It's about 6 ft wide and the roof is 20 ft high. On the right-hand side of the corridor, as you walk silently down it, are two blue doors. High on the wall above the doors are red illuminated signs saying 'Command Enabled'. They lead to the heart of the manned spaceflight programme, the Manned Operations Control Room, the 'Mission Control' you see on television. Down the left-hand side of the corridor are three doors. Each one is almost always open. Inside is a room about 20 ft square. The wall opposite the door is set the length of the room with glass, and through that you can see another room, identical in size. Running through both rooms are rows of consoles. They are grey, like the walls and the floors. During the months between missions, these support rooms, as they are called, are more often than not empty. But the consoles in them are always at work. Their screens are filled with row upon row of figures, computer data. And if you listen carefully you can just hear a ghostly conversation coming from the loudspeakers buried somewhere in the walls.

'Flight, Ecom. We're getting uplink on C.'

'OK. Can we get some more numbers on that last phase?'

'Yes, Flight. And could we get Capcom to tell the CMP to try circuit breaker 3D, and if that doesn't do it we'll come up with another procedure for the data rate in a minute.'

The figures on the consoles change. Another whispered exchange begins. You're listening to a Sim. NASA speaks abbreviations. This one means Simulation. A trial run of some part of another infinitely detailed spaceflight to the Moon. Before any Saturn lifts off on the white ball of fire that looks so awe-inspiring, and arouses the hidden desire in all of us to see blood and flames and explosions, every second of the flight ahead has been rehearsed dozens of times. That sounds like exaggeration. It's not. The simulations, coming in pale colourless whispers out of the blank grey walls of the Mission Control building, contain all the blood and flames and explosions that NASA hopes will never really happen. This is where the crew and hundreds of men who support them on Earth get their last chance to live through disaster, even, in the case of the astronauts, get killed – on a reel of computer tape. Of course you can't plan for every eventuality – look what happened to Apollo 13. But you can run through the most obvious ones. That leaves only about a million or so unrehearsed, if you count all the tiny components on board the spacecraft that *could* fail, and set off a chain of events leading to destruction. The one on Apollo 13 was less than a half-inch across. Still, everybody tries.

As you stand in the grey, air-conditioned room, listening intently to the unemotional tones of the scientists, engineers and astronauts, sitting themselves in grey air-conditioned rooms all over America, there comes over you a strange sense of being out of your time. You feel that, if you went across the corridor and opened one of the blue doors, you'd step into the beginning of the next century. Not that the rocket, and the Moon mission or the whole NASA manned spaceflight complex are in themselves futuristic. It's the voices. There is never the slightest sense of uncertainty in anything they say during an exercise. While there are three men on their way out into space the Capsule Communicator – the astronaut who sits in

Sea Breezes. The Indian subcontinent and Ceylon from 600 miles high. Note the great U-shaped cloudline around India. This shows the weather system that so often provides blue skies at the beach while inland and out to sea there are clouds. As the wind from the sea crosses the coast it rises, to form clouds over land. Out to sea higher air drops to take the place of the air mass moving inland. This too causes clouds.

the very centre of the Operations Room during a mission – will actually joke and pass the time of day with the crew. He will do so in strict obedience to Mission Rules. One of the Capcoms once said that his job specification was matching the mood of the crew and responding accordingly. The rest of the Control team act as one with the machines that serve them. It is this 'man-machine interface', as NASA would call it, that you feel conscious of when you stand in the Operations Room. The men here appear to have almost total control over the environment in which they work. Perhaps it has something to do with the fact that the simulations are in a sense so detailed, the mission rehearsal so totally related to the behaviour of engines, dials, instruments – as indeed is the actual flight for which they practise – that in reproducing the internal working of the spacecraft in all its complexity the Simulators are creating reality itself.

It is this feeling that the Controllers can mould and change a flight, irrespective of any natural intervention by welding their identities with machines to bring about that change, that is futuristic. Here the rest of us are, uncertain whether or not to take a raincoat when we go out, while they can compute and control the world of three men at a distance of 250,000 miles. Of course in each separate part, at each level of technical activity, what they do is understandable enough. It is the totality of what they can do that whispers the end of the century at you out of the walls in that empty room.

In another one of the white concrete and tinted glass buildings at Houston, the Simulators go even further: in the spacecraft training simulations they actually create a picture of the lunar surface for the pilot to land on. Every action to change the direction of the ship is transmitted by computers to a camera scanning a relief model of the landing area. The camera swings, and inside the simulator cabin it is difficult to lose the sense that you are actually looking out on the Moon as you fall gently to a touchdown. The training in this controllable reality is so intense that Pete Conrad, the Commander of Apollo 12, said the landing was just like a simulation. But what does such deep-level preparation do to the astronauts' sense of reality during a flight? Conrad said he felt no sense of excitement at being on the Moon – only a feeling that he belonged there. It is a temptation to react to that by saying that in one sense the entire episode may never really exist for the astronauts. The mission cannot give them any sensory experience that their masters have not already given them on Earth. Indeed they take Earth with them – in the form of their spacecraft and suits, and the life-giving oxygen within them.

Dave Scott, Commander of Apollo 15, once said that a flight is nothing so much as minutes and seconds going by, each one related to a task that has to be done. Perhaps in one way that's all a flight to the Moon is: the completion of a period of time. Direct contact with the new world they explore is denied them. They never touch, smell, hear or feel what lies outside their windows. And without that how much more real is the view than the simulations played on the television screens back on Earth? There is no way to make what lies outside any more real. Removal of helmet or gloves would mean instant death. In years of talking to crews, it has been impossible to get from them a reaction to the flight that would compare in any way with the raw sensory imprint that any earthbound adventure leaves on even the most experienced human being. On the other hand it may be that such excitement might increase the adrenalin load in their bloodstream, and that could lead to a quickening of the breathing. That in turn would cut down the time available on air supply as it went at a quicker rate. So perhaps the excitement of reality lessens the efficiency with which the mission can be carried out.

The astronauts themselves are intensely efficient people. They rarely make expansive gestures or talk loudly. At parties where anyone from outside the astronaut corps is present they are like shy wallflowers. If you approach them they are extremely polite, almost disarmingly so. Just how disarming you discover when at the end of the evening you find out how little you've learnt about them.

The biggest room in Space – Skylab's workshop (see p. 26).

Somebody has given them all very thorough training in talking to outsiders without really saying anything. It is in this way – how they communicate with you – that most astronauts are similar. In essence the message you get from fifteen separate individuals, before they head for the Moon on five different occasions months apart, is the same: 'No particular part of this mission is any more dangerous than any other. Every part is essential if the mission is to succeed. At launch I guess I will just be feeling it'd be good to be on my way. The landing on the Moon demands piloting skill, but we feel pretty sure we know as much as is necessary to take the danger out of it. Yes, going to the Moon is necessary because mankind expands his horizons. If he doesn't no progress is possible. No comment on what any political figure has said against the space programme. We're looking forward to a good flight and to helping the scientist planners as much as we can. Thank you.'

That middle sentence, 'We know as much as we need . . .', may hold the key to what they are. Never have so few known so much about so little. Each group of three Apollo crewmen are more intimately connected with their spacecraft during the last two years of training for a flight than they are with their wives. Jim Lovell thinks that after the months of living in the cabin simulators you develop a sense of contact with the machine, an instinct for knowing when something isn't functioning as it should. But perhaps the clearest expression of that knowledge of their little world, and the effect of such knowledge on them, was given by Neil Armstrong at the Press Conference before Apollo 11 lift-off. He said that fear of what might happen to a man on a spaceflight was a matter of how much he didn't know about what might happen, and since all of the crews knew in detail what would happen in the event of an emergency, the question of fear didn't enter into it. Fear, all the astronauts say, depends on ignorance. Inside their spacecraft – that includes the suit, which is in itself a tiny spacecraft propelled by muscle – nothing can happen they don't know about. Outside it, there lies not so much the unknown as the unreal. If you ever contact the outside you're dead so fast there is no time for it to exist for you, no time for fear. It is some comfort to the layman that, in the company of such superhuman utterances, he can fall back on disbelief. And yet again there is the feeling that we might be wrong. Perhaps they really do believe what they say. Perhaps they are therefore a breed of men that will not become common for another hundred years, when the rest of us have learned to control our environment to such an extent, and the era of Natural Man comes to a close, as he and his machines combine to produce a new composite being, separated from the rest of the animal world for good.

There is, perhaps, in spaceflight what the futurologists would call a scenario, a possible sequence of events that may happen during the next century. The Earth is our spacecraft. Already, as you can see in other parts of this book, we are beginning to understand how its engines work, where to look to see the information we need on the instrument panel, how much life-support reserves we have. But there the resemblance ends. The Earth has no goal at the end of its flight, and if we have an emergency there is no base to return to, no support team to solve our problems and get us out of it alive. There is nowhere else to go, or if there were, no means to get the crew there. And we do things to the machinery of our spacecraft that no astronaut this side of insanity would do to Apollo. It may be that the greatest danger is that we will adopt the Moonflight scenario for our future and attempt to control what happens in our spacecraft. Already we have our own simulations in the form of very primitive mathematical models of the biosphere: the tides, the weather, urban development. We are just beginning to be able to predict what will happen if we tinker with the machinery. But even if we do succeed, and come through unscathed to the day where the simulations are put into effect, and the spaceship Earth comes totally under our control, what then? What will we be? Will we all be astronauts, each one with a checklist for living, never deviating from the mission rules for fear of upsetting a world-wide flight plan?

The answer to that question may lie along that grey corridor in Mission Control, for as you reach the end, just opposite the last room on the left, you see that the wall needs painting. It's like discovering that the idol has feet of clay.

NASA once had plans for ten landings on the Moon. When Apollo finishes, in late 1972, there will have been six. And there are no dates for a resumption of the Moonflights. All over America workers on the space programme have been laid off. The budget has been cut to the bone. At the once-busy Press Office in Cape Kennedy, rows of offices and desks stand empty, because with the cutbacks has come a great fall in public interest. You can see it in the Auditorium at Houston where they hold the pre-flight briefings for the press corps. In 1969 you had to get there early to be sure of a seat near enough to the astronaut-gods to ask a question. Even the hardened newspapermen craned their necks to get a glimpse of Armstrong, Aldrin and Collins in the flesh. They spoke from inside a plastic tent blown with sterile air – their words like voices on some Olympian wind. They were untouchable. The rest of us were literally too dirty to come near them. No germ could be allowed to contaminate the bright perfection of their last days on

Earth before the great journey. Today you can take your kids to the briefing, as you would take them to the zoo. The Press is still there – a small gathering of the faithful. The crew no longer take you through the mission like priests revealing mysteries to the chosen few. They express appreciation that you are there at all, and ask for questions. Everybody tries hard to think of one, to give the situation some semblance of occasion. The wheel has turned full circle. Once the astronauts said they were just ordinary test pilots flying a different kind of plane, and nobody believed them. Today they are imbued with a great sense of purpose, and nobody will listen. Tomorrow the death of an astronaut as he burns up on re-entry from a space station will excite no more comment than an airforce pilot who hits a mountain with his fighter. Those of us who saw and heard the first few landings on the Moon may have seen the only brief moment of glory that spaceflight is likely to have for the next hundred years. The proposed landing on Mars has already been devalued by Apollo. There is no way we can appreciate a visit to Jupiter any differently – it's just farther away. The next great adventure lies perhaps hundreds of years in the future, when we cross to the stars, if indeed we ever do.

One of the Apollo 14 crew said before he left that he couldn't conceive of spaceflight being terminated. It would be, he said, like deciding we had gone far enough without ever knowing how far we could have gone, and mankind, he asserted confidently, wasn't like that. Yet it may very well be that lack of a goal which *could* cause mankind to prove him wrong. It is difficult to call to mind any example of exploration by Man where he could not see or conceive of the end of his search. Even if that end were totally unknown, it was on a scale that he suspected he could encompass with his mind. In that sense nothing has ever been beyond the measure of Man. But perhaps the manned exploration of space is the only truly open-ended search that we have ever embarked on, and perhaps it will be the only one we ever give up. No doubt one day towards the end of the century, a lonely space probe will send back across the orbits of the inner planets a ghostly message that there is aluminium on the surface of Mercury, or that Pluto's ice mountains are made of frozen formaldehyde, and somewhere a technician will pass the message on to a scientist who cares. But will anyone else? Today the Apollo flights to the Moon excite only those of us prepared to become involved with the findings of an alpha-particle analysis of some part of the chemical composition of the lunar soil, or the impregnation of a piece of foil by the solar wind. The briefings at Houston include long, poorly-attended dissertations by scientists who talk at great and informative length about their special interest in the mission.

The great adventure is over. And a good thing too. Now, after 10 years of show-biz publicity and death-or-glory exploits, we can perhaps get down to the humdrum business of using the tools developed to go into the wild black yonder for something of more immediate benefit. The next few years of manned flight may not make television spectaculars or front-page headlines, but as the men of tomorrow pilot the ships to orbit with the space stations endlessly circling the globe, their scientist passengers will need no roars of public acclaim to help them get on with the work, or to get the money to pay for the work. Much of it will already have been paid for, thanks to what somebody once thought was a race for the Moon, and the beginning of man's conquest of the galaxy.

When you talk to the scientists briefing the newsmen at Houston before these latest Apollo flights, or attend one of the 'Rock Conferences' in the sand-coloured city conference centre in downtown Houston, which is actually 22 miles north of the Manned Spacecraft centre, you get the definite impression that many of those involved in these aspects of manned flight are almost relieved at the fall in public interest in the missions. That is not to say they would wish to keep the rest of us in ignorance of what they're doing, but the past few years of

Skylab control centre —
the Multiple Docking
Adapter (see p. 24).

excitement and exaggeration attending the flights have, in a sense, obscured the real value of space exploration. You feel that as they group round blackboards, scribbling incomprehensible formulae to shouts of approval or cries of 'charlatan' (it's happened) the scientists are pleased at last to be able to present the facts in their own terms, to be taken up or ignored by the public at large. How many of them have refused to be interviewed because what they have to say cannot be condensed to a 2-minute film report or a quick paragraph. Now that the circus is over, they can return to the anonymity most of them prefer. This is not to denigrate the risks taken and the courage displayed by the pioneer Moon-landers. The Apollo missions, when they finish, will have brought back enough material from the Moon to keep laboratories all over the world busy for a decade. But their exploits have confused people as to the real value of the tremendous technological feat of putting men on the Moon.

NASA knew as well as anybody else that there would be a great sense of anticlimax after Apollo 11. When all the excitement and cheering died down, considerable damage had been done to NASA's chances of getting the money they needed to go on using the technology they had developed. That, it appears, is the real reason the last three Apollos – 18, 19 and 20 – were cancelled. As the heat subsided, and the voters came out of their patriotic euphoria, they saw the size of the bill and called their political representatives to complain. Even if America had not been heading into a recession at the time, it's a safe bet that the purse strings would still have been tightened.

All the public had been able to get for their money was a low-quality picture from various lunar landing sites. It could be said that even if the pictures had been of superb quality, the same would have happened, because inevitably Apollo became a television programme, and as every television producer knows, if you want to keep your audience, you have at least to vary the plot. The ratings on the TV Moon shows fell faster than almost any other series the American networks have ever aired. And since there was no other way for the public to participate in the lunar adventure, they reacted like any audience to a repetitive series. They switched off. NASA's budget almost halved, and the order went out to stop production on the giant Saturns. With only a limited number remaining, at least two had to be earmarked for the project planned to follow Apollo, a mission that needed the $7\frac{1}{2}$-million-lb thrust only the Saturns could provide. Without them the project, known as Skylab, could not exist. So to the considerable chagrin of the Moon scientists, the lunar missions were curtailed, and Skylab was safe.

With Skylab comes the virtual end of space exploration as we conceived of it back in the early 1960s. Although Moon visits will be resumed in the 1980s, the blast-off into the unknown regions beyond the orbit of the Moon looks like remaining the property of the science-fiction writers. There are plans on paper for an American manned landing on Mars in the late 80s or mid 90s, but unless a real or imagined political reason for the flight emerges, as was the case with Apollo, then it is difficult to see the money being provided. A manned flight to Mars is, after all, not just a matter of building a spacecraft and life-support systems capable of taking six men there and back. In order to mount such a venture you have to spend so much on designing the machines to build the craft that it is uneconomic to limit the production line to one masterpiece. So even before you begin to cut metal you have to commit yourself to several flights. That means a full-scale Mars exploration programme of the Apollo variety.

Perhaps the best way to explain just how valuable Skylab will be is to fly on it. The day before you go, you'll watch the Saturn V take the orbiting laboratory into space. The countdown in the Firing Room is the same muted event it always is. The Launch Director's desk – a row of consoles running the length of the room – sits about 12 ft above the rest. Below you the dozens of launch officers sit hunched in front of their own consoles, watching the figures change on the screen in front of them. Each one of them is at the receiving end of a communications network. At the other end, in other rooms throughout the Launch building, hundreds of individual engineers watch their own particular part of the Saturn as it sits hissing on the launch pad 3 miles away towards the sea. Outside, the last few minutes pass before lift-off. The giant umbilical tubes are still attached to the Saturn's side, topping up the supercold hydrogen and oxygen, replacing the gallons that boil away every minute in the warmth of the Florida sunshine. The structure creaks as the breeze sways the rocket very slightly, and dozens of tiny clicks and taps sound, relays closing, valves operating in the final stages of preparation for lift-off. The engineers can hear none of this – nobody can. The launch pad is deserted. But on the consoles in the Firing Room each click, every creak of metal shifting, appears as a series of figures on the screens. As each group of figures appears, each man's electronic jigsaw fits together. He touches one of the square buttons on the right-hand side of his console. A light flashes on the same button on a console in the Firing Room. A brief exchange, and the launch moves one minute pace closer. When all his reports have come in, the officer at the console calls the Launch Director. He knows he will be the fourth or the second or the tenth to do so, because his report comes in a particular part of the final sequence. When he gives it, he says two words: his designation, and the word 'Go'. As you sit by the Launch Director in the last few seconds before ignition that's what you hear over and over again: 'Go. Go. Go. Go.' When it does go, only the uninitiated actually turn and look out of the giant louvred window to the rocket, shimmering in the haze, to wait for the sweep of the flame and the roar, and the infinitely slow rise on the white ball of fire. That's only the end-product of the dancing figures on the console screens, the spectacle for the crowds to watch. One of the launch officers once said that when he actually saw a lift-off, it was an anticlimax. The real surge of excitement lay in the patterns on his console, the real beauty in the preordained precision with which they changed.

And so the laboratory lifts into orbit, a specially rigged-out Saturn V third-stage tank, carried up inside a protective nose cone. You leave the Launch Room, and return to the sparsely furnished quarters a couple of miles away back up the road to make ready for your own launch. Meanwhile, as soon as the launch pad is cool enough, the engineers move in to prepare for crew lift-off next day, on board the smaller Saturn 1B, unused for spaceflights since the launch of Apollo 7 on its Earth orbit mission back in 1968. The way this smaller rocket is serviced and launched from the Saturn V pad is quite ingenious. Because the gantry carrying all the fuel and electrical lines to the V will be used, the 1B will be launched sitting on top of a pedestal, which itself is sitting on top of the launch pad. That way the access arms on the gantry will be level with the sections of the 1B that need servicing during countdown.

After a good night's sleep, aided if need be by drugs that leave no after-effects, you wake on the morning of launch. Breakfast is always more or less the same for crews: steak, eggs, orange juice. This is not because all those who go into space develop similar tastes during training. It's for the very good reason that you need to take as little roughage as possible up with you inside your body. It cuts down the chances of your having to defecate inside your spacesuit. But just in case, one of the last things you'll do before putting the suit on will be liberally to smear your buttocks with a Vaseline-like substance. Then if you do excrete, though most of the faeces will be absorbed by the special cotton-wool drawers you wear, the rest won't harden and irritate. Getting into the spacesuit itself is a matter for contortionists. It's tailored to fit you, and what that means is that it's specially hand-made in a small, very English-looking town called Dover, in

Delaware, for a cost of nearly £50,000. In Earth gravity you need help getting into it. It's a strange sensation finally to put the plastic 'goldfish bowl' headpiece on. It comes so close to the face that for a moment you feel claustrophobic. Then the cool hiss of the portable box-like air system you carry into the spacecraft calms you down.

Walking down the corridor to the waiting truck is something many of the men already used to launches want to get through as quickly as possible. The suit is quite cumbersome – you waddle towards the truck waiting to take you to the pad, knowing that in a few hours it will weigh nothing. At the moment you can hardly turn your head without moving your whole body. But apart from the weight, the suit is extraordinarily comfortable. You've set the temperature you want your body to be at, and whatever your exertions, the underwear, ribbed with hundreds of tiny tubes carrying water from your air-supply pack, will take away any heat you build up. Rusty Schweikart, the crew member of Apollo 9 who was so unfortunately nauseous during his one and only flight in 1969, described it as being like standing under a cool shower.

The truck finally delivers you and your shower to the pad when the countdown has some 3 hours to go. As you ride the lift up the side of the service gantry to the capsule, if you crane your neck, suit permitting, you can just see the flame trench below the Saturn, still cooling from the effects of yesterday's launch. The angled floor of the trench directly below, that takes the full force of the blast and deflects it sideways out and away from the rocket base, glitters in the morning sunshine. The reflections come from the fused glass-like surface. Under the tremendous heat generated by a Saturn trying to get off the pad, the surface of the trench floor runs liquid for a moment, then blackens and hardens to a smoothness like black ice.

At the top of the gantry there's a short walk across to the white room, the sterile white windowless box – the last thing you see before the hatch is closed. A couple of technicians help you through the small hatch into the capsule and onto your couch, on your back, staring upwards at the dozens of dials and lights on the instrument panel. In time, the protective nose cone, with its escape rocket mounted above it to carry the capsule up and away on a tumbling path should anything go wrong up to a few minutes after launch, comes down over the hatch window, and you and your fellow crewmen are alone in the blackness. Everybody says, without exception, that during these last minutes of countdown there is so much to do, so many instrument readings to complete, so many checks to go through with the controllers 3 miles away in the Firing Room, that you think of nothing but the work at hand. Well – nobody who's flown can remember having thought of anything. But in the very last 10 seconds, in the words of Jim Lovell, 'you hear the engines rumbling up, there's a very slight amount of vibration, you count the seconds, you feel the restraining arms on the base of the rocket go back – and you're either gonna go, or you're not gonna go'. It's supposed to be a very smooth ride. On occasions bad weather has caused buffeting, but in general the climb is gradual, as, with the windows still covered by the nose cone, you ride up through the first stage of the flight to orbit.

At just after 3 minutes into the climb, the escape tower system is jettisoned, and you are flooded in sunlight streaming in through the windows. Earth is below and behind, because you climb to orbit on a long curve that ends up heads down. The second-stage engines cut in with a shove that sends you forward against your restraining straps. Long ago, as the rocket passed through the speed of sound, the roar of the engines on lift-off has gone. The place is silent. Finally the Saturn is in orbit, and the computer, just left of centre on the displays in front of you, begins to work through the early stages of the rendezvous ahead with Skylab, in flickering patterns of light figures. The only steady one is the group of three at the top of the display – the Programme, the Verb, and the Noun.

These literate-sounding designations refer to the orders you or Mission Control at Houston (they took over the flight from the moment the rocket's tail cleared the launch gantry) fed into the computer. For crew input there is an extremely simple keyboard, set in plastic squares that depress slightly when you touch them, and light up to remind the operator that he's punched that particular key. On each key is written a number or a word: VERB (the information or action required from the computer), NOUN (what the computer is to do with that information: display every second, store for retrieval when needed, etc.), the numbers 0 to 9, CLR, STBY, KEY REL (release), ENTR and RSET, and + and −. It is with those 19 plastic keys that the mission is flown. The hand controls that activate the tiny directional thrusters on the outside of the spacecraft are strictly for small-time manoeuvres. They're used during the final moments of docking, or to angle the spacecraft to get a better picture out of one of the windows.

In the midst of all this computer-gazing, you discover you're weightless. Going into a zero-gravity state, as it is more accurately known, is either one of the most exciting, or one of the most unpleasant, experiences you may ever have. Which of the two it will be for *you* will have been discovered months before lift-off, in a training flight from the Wright Patterson Air Force Base, at Dayton, Ohio. The prospective candidate is taken to a height of about 30,000 ft in a 4-jet military transport plane that looks rather like the conventional Boeing 707 used by airlines. This plane, however, has no seats. The entire fuselage is empty, save for thick padding that covers every inch of the interior. At altitude the plane begins a dive at an angle of about 45°, building up speed to over 500 mph. At about 22,000 ft the pilot pulls the stick back sharply and the nose of the plane comes up. At the same time he cuts the power on the engines. As the aircraft pulls out of the dive, the wings flex downwards so far that if you look out of the window you can see only the first few feet. You can't believe the structure can take the load. The whole aircraft creaks and groans under the tremendous strain as it responds to the sudden change of direction. You are pinned to the floor, the blood draining downwards from your face. You panic – no aircraft can take this punishment – the wings must come off. But there is nothing you can do, not even move a finger. The whole thing lasts perhaps 15 seconds – it seems an eternity. Then suddenly the manoeuvre is complete, the plane's nose lifts steeply, and the aircraft begins to follow a high-curving parabola. The crew describe it very succinctly as 'going over the hump'. During this manoeuvre the forces acting on the plane exert a pull upward exactly equivalent to the pull of gravity downward. As the parabola begins all noise stops, and in the eerie silence that follows, you rise off the floor. There's an instant's shattering panic – you're falling. You throw out a hand to save yourself, and you begin to spin slowly in mid-air. Then, a second later, the panic is gone, to be replaced by the most extraordinary feeling of delight. There's no other word for it. Your throat fills. You become light-headed and enormously happy. Most people who go through this first period of weightlessness do so with a foolish grin on their faces. So do you. You try to speak, and all you can manage is a squawk. Almost instantly the feeling of being any particular way up disappears. As the body spins, or hangs 'upside-down', it merely appears that the rest of the world is moving. There is quite literally no physical sense of up or down. Only a great feeling of freedom – movement in any direction is effortless. If you've ever dreamed you could fly, this is what it's like. It is, as one astronaut said, the second most exhilarating thing anyone could ever do. If, however, you suffer from motion sickness, it may be the most unpleasant thing ever to happen to you. Vomiting can be almost instant, and in space very dangerous, for the particles hang in the air, to clog the breathing passages and possibly asphyxiate the sufferer, certainly if he is wearing a space helmet at the time. The body and its functions feel so different that they are the subject of constant awareness.

In the relatively confined space of an Apollo capsule some of the private things your body does can no longer be private. The worst of all, according to one astronaut who wants to remain nameless, is going to the lavatory. After some hours in space, sooner or later it has to happen, and when it does, in spite of the air purifiers, the interior of the spacecraft becomes almost unbearable. 'Then,' he says, 'in time you get used to it, until it's somebody else's turn, when the whole extremely unpleasant experience happens all over again. In time you accustom yourself to it all. But the man I feel sorry for is the poor son-of-a-bitch who opens the hatch when we get back from a particularly long flight.' So as you coast up towards Skylab, don't feel ashamed if you're hoping you make it to the lavatory-equipped orbiting laboratory before you have to go through it in company. Of such things are the greatest adventures made.

Almost immediately you reach orbit the spacecraft's engine fires, thrusting you back in your couch, to lift the orbit to match that of Skylab, riding over 300 miles up. There is little time to enjoy the view, as the computer pulses the tiny directional jets to make the necessary corrections to the trajectory that will bring Skylab in sight perhaps an hour after lift-off, and you check the instruments almost without pause. These trajectories are not the mysterious, intuitively-judged things that are, say, involved when a pilot brings his aircraft in for a landing. Once in orbit an object will swing round the planet in the same way until its path is deliberately changed. But since being in orbit depends on a delicate balance between your speed and the gravitional pull of the Earth, you are very much in the position of the chestnut on the end of the string. If you want to fall back towards the Earth, and lower your orbit, you simply fire the engines on the spacecraft into your direction of motion. This retrograde firing will reduce speed, and the orbit will lower. The opposite is true for lifting the orbit higher. This is why the simple-looking business of rendezvous and docking is so extremely difficult. As two spacecraft come together, one of them will at times have to catch up with the other, or slow down to avoid overshooting. The trouble is, as you've seen, extra speed lifts, less speed lowers. So as you hurry to the target you rise above it, and as you put on the brakes you drop below. What the computer does is achieve a compromise, by altering the orbit of your spacecraft, so that you fly an elliptic path round the Earth, which intersects with the orbit of Skylab at a certain point. The intersecting rendezvous orbit can be set with extraordinary precision, since the orbital behaviour of your capsule is so un-varying. Frank Borman said once that you could set a clock by the regularity with which a spot on Earth passed under you. 'Every hour and so many minutes,' he said, 'we'd look out of the window and there was Arabia again.'

The last few miles of the rendezvous with Skylab are the worst. They certainly take the longest. The orbiting laboratory will have been visible ahead for some time, sitting motionless in space like one of those trick drawings that deceive the eye through misuse of perspective. As the sunlight reflects off the white surfaces on the laboratory, and the unlit parts merge with the blackness of space, it takes you a while to sort out the shape. The Commander of your Apollo is meanwhile going through station-keeping – what the crews call the waltz. Sitting a few feet apart, he rolls, pitches and yaws the spacecraft until the Apollo probe is aligned exactly with the docking tunnel in the end of the cylindrical section sticking out from the laboratory, containing two such tunnels. They're known as docking ports. This alignment must be absolutely exact. If not, either the tunnel or the probe will meet at the wrong angle and possibly damage each other. The ease with which you float there, seemingly within touching distance of the side of the laboratory, is dangerously deceptive. Those astronauts who have flown Lunar Modules back to rendezvous with the mother ship have called these last few moments of docking the most hair-raising of the whole mission, because what you are floating about in is several tons of inertia that will go straight through anything it hits at any speed. Steering is a strange business for anyone used to flying in atmosphere, because once you pulse one of the directional jets to put the spacecraft in motion, she will go on in that direction until the opposite jet pulse is fired. There is no slow-down, no change of direction as you move. If you are spinning and swerving at the same time, the process of working out which set of jets to pulse in order to bring you to a stationary position any particular way up at a certain spot is, to say the least, a complex one. And in a weightless condition you have no sense of 'down' to help you.

As you cross your fingers and leave the details to your pilot, you look out across the intervening space to Skylab. It blots out the stars – a giant tapering cylinder with what look like broad wings near one end and four helicopter rotors at the other. The wings and rotors are in fact solar panel arrays, studded with a mosaic of thousands of light-sensitive cells that turn the sun's light into electrical power for the entire structure. Those at the front provide power for the main systems, the four near the back feed the Telescope Mount and its subsystems. It is in under these panels at the back end that you and the Apollo capsule nose the last few feet to the docking tunnel like a fish gliding under a frond. You hear the thump as the nitrogen bottles release their gas to operate the claw-like device on the end of the probe which secures it to the tunnel, and Apollo stops. You're there.

Once the latches around the mouth of the docking tunnel have snapped shut, the probe itself folds up and comes back into the capsule. Now on a small scale comes the inertia problem again. As the steel probe, looking rather like a half-closed umbrella, floats around while it is being stowed great care has to be taken not to bump it against the instrument panel, activating any of the dozens of switches. Once it's safely under one of the couches, the tunnel cover is removed, and you and your fellow crewmen float through into the first room of the laboratory. It's little more than a narrow, 10-ft-wide tunnel, packed with consoles and handrails. If the thought makes you feel claustrophobic, this is where your fears are allayed. Everybody who's been into space says that weightlessness seems to more than double the space available. If the 'floor' is cluttered, you float above it. The tightest corners become easily accessible.

This room is the Multiple Docking Adapter. Immediately to the left of your entry tunnel you see another tunnel cover set on the end of the 'back-up' docking port. This is to allow another spaceship to arrive at Skylab, in the event of an emergency rescue. To the right of you is a large console covered with dials, and what look like radar scopes. This is the control console for the Telescope, mounted on the outside of the Docking Adapter directly above. On the radar scopes, which are actually television screens, will be projected the object for whose study Skylab is primarily designed – the sun. Above your head as you sit at the console, outside the curved wall of the Adapter, is the 10-ft octagonal canister housing the electronics for all the Telescope experiments. Mounted on top of the canister is the 'dish' carrying the necessary sensors. Astronomers have always wanted a view of the sun unobstructed by the atmosphere. This is it. And it is for this reason that Skylab will always fly in such a position as to present the sensor dish directly towards the sun.

The rest of the Adapter is filled with a seemingly chaotic number of electronic equipment packages and rows of pipes, as well as storage cupboards – everything fitted with handrails. It's only as you float in mid-air and let yourself slowly rotate that you see order in the chaos. What appeared to be a luggage-rack without purpose turns out to be a footrest platform for the crew member working on a set of small consoles and cupboards. This is the great virtue of the zero-gravity state – every inch of wall, floor and ceiling space can be utilised for experimental equipment.

The aluminium ring top cans containing Skylab food. Heat and eat (see p. 28).

Two crewmen Train on a Skylab mockup of their living quarters (see p. 27).

WORKSHOP MOCKUP CREW QUARTERS

ELECTRICAL OUTLETS

FOOD WARMING/ SERVING TRAY

WARDROOM

REFUSE DISPOSAL

NO STEPPING O

At the other end of the 13-ft-long Adapter, you float through a hatch into an area even fuller of equipment. This is the Airlock module. Slightly narrower than the Adapter, it houses all the controls and equipment for activating and operating the entire laboratory structure. The walls are packed with consoles, dials and switches. Between each bank of instruments is the usual luggage-rack foot platform. There are also windows. This is the heart of Skylab. The minimum-power lights that greeted you as you floated through the docking port are replaced now by high-intensity strip lighting that flicks on all over the Skylab as the Commander of your flight moves from console to console bringing the station to life. Needles move, lights wink, and the place begins to fill with air from storage bottles set around the Airlock Module itself. Before anybody moves on further into Skylab, every switch and button must be set and rechecked to make sure that the life-support systems, the heaters, the air-conditioning, the scrubbers removing carbon dioxide from the air, all are functioning properly. Only then, still moving awkwardly in the white Apollo spacesuits, do two of you move towards the small latch at the inner end of the Airlock Module. It's about 2 ft 6 in. wide, and as you pull yourself carefully through it, you find yourself inside the biggest room ever to go into space. Three times larger than the Russian Salyut station launched back in 1971, the entire shell stretches out for over 10 ft on each side of you, and for more than 48 ft above your head, as you ease in through the entry tunnel.

Just to one side of the tunnel is a pole running away through the centre of the main laboratory area. You glide 'up' the pole, carefully propelling yourself along with one hand. Again all around the walls are electronic packages, strange-looking canisters and benches, and laboratory equipment of all kinds. By each cluster of equipment are cupboards and storage nets to hold tools and materials. No luggage-racks here, since beside each area, set on the wall as you pass, are odd sections of criss-cross metal mesh. You are to discover the significance of that mesh later. About 20 ft along the pole there is a section of the same mesh stretching all the way across the laboratory. The pole goes through it via a 3-ft hole cut out of the centre. You realise immediately that this is the floor of the workshop area of Skylab, the area you've just come through. On the other side of the mesh is another space, about 8 ft deep, and then another mesh floor. Reversing your body, you come to a gentle halt with your feet on this second mesh, and you are standing on the floor of a circular room, crowded with instruments and what looks like exercising machines. To one side you catch sight of vertical hammocks and notices saying 'Waste management section', 'Food', 'Freezers'; you're in the Ward Room, home for the duration of your stay on board.

By now the crewman left behind in the Airlock Module indicates that the pressurisation and heating is complete, and with considerable relief you take off your suit. Although even the most advanced suits are pretty comfortable, the one thing you can't do in them is scratch yourself. That's the first thing you do when you take one off. You may not need to scratch yourself – but you do. It's a kind of psychosomatic scratch – to prove that you can. Once the suit is stowed (and the latest types fold up into a suitcase about the size of a businessman's overnight bag) you get the underwear off and head for the shower. The protective Vaseline substance between your legs wipes off easily enough. You step – or rather float sideways – into the shower cubicle. It's unlike any shower you've ever taken. It may be one of two types: either as the water drops rush out of the shower an overhead fan starts up, sending a stream of warm air at high speed towards your feet, so that instead of floating about, the droplets are literally blown over your body and sucked out through a grille on the floor; or if you only feel like a quick wash all over, you may pick up a device looking like a sponge on the end of a hose. A twist of the taps brings a trickle of water and soap oozing out to cleanse the skin. Another twist, and you're suddenly patting your

Lunch 300 miles up. The Skylab Kitchen. "Cooker" has zero g seat-sticks (see p. 28).

wet body all over with a tiny vacuum cleaner, as the moisture in the sponge picked up from the skin is sucked back up through the hose and away to waste-containers. Wherever the water ends up, it's put through purifiers for reuse later. Nothing on board is wasted. Even the air you breathe is being cleansed, cleared of all carbon dioxide, and recycled through the fans set along the ducting tubes running the length of the laboratory.

The principle of wasting nothing is in operation even when you finally get to the lavatory on board. Urine goes down a small hosepipe to be collected for analysis on board. Faeces are collected in plastic bags. The upper edges of the bag are adhesive. You peel them apart and stick one to each buttock. When you've finished, instead of toilet paper, special tissues impregnated with a germ-killing substance are used. You reseal the bag using the adhesive edges. In a tiny pocket on the side of the bag, separated by a thin membrane, is a germicide pill. Pushing on the pill breaks the membrane, allowing the pill into the main bag with the faeces. You then knead the bag until the pill has broken up and mixed thoroughly with the faeces, turning it blue. At that point the faeces have been thoroughly cleared of potentially gas-producing bacteria, and the contents can be stored for later analysis. It's not exactly a luxury toilet system, but all the astronauts infinitely prefer it to the absorbent nappy-like drawers used inside the spacesuit! At least, they say, on Skylab you have a modicum of privacy for the first time in space.

Later on during the day comes the first meal. The kitchen looks more like a small computer terminal. Around the walls there are conventional freezers that any housewife would recognise. They contain the food, either vacuum-packed in plastic bags or held in tins with the familiar ring-pull opening system used on beer cans on Earth. Liquids, like fruit juice, come in powder form, packed in bags with a tiny opening for the insertion of water-pistols used to rehydrate the contents. The food in the cans is totally unlike what you have come to recognise as space-food. The cans fit into the object that looks like a computer terminal in the centre of the kitchen. A central column rises from the floor to waist height. At the top three box-like compartments project sideways, one for each crewman. You lift up the lid on your box and discover eight circular holes set into a smooth surface. Each of these holes will take a can of food. Switches on the outside edge of the box, marked 'Left', 'Front', 'Back', 'Right', turn on heaters coiled round the inside of the holes. Turn on the heat for the desired number of minutes, and the precooked food in the can is ready. Peel back the can top, and eat: meat, stew, vegetables, puddings – whatever you feel like. And if the idea of eating while standing up doesn't appeal to you, hook your feet around the central column, and eat while floating in a sitting position, or any other you fancy.

Late in the evening of the first day you report a summary of the day's events to Mission Control. By this time the old-fashioned Ground Elapsed Time of the Moon Missions will no longer be in use, except for computer record purposes. You keep the hours of the day according to Cape Kennedy time. Finally you turn in for the night. The sleep alcoves on board are in the Wardroom, to the left of the kitchen, and as you zip yourself up in the vertical hammock you prepare for the strangest sleep of your life. Even though the hammock fits snugly enough there is one vital thing missing. You can feel no pressure on your body, as you would in bed back on Earth. Apparently it's a deeply ingrained signal the body needs, to know that it is lying down, and some of the men who've flown in space before, both Russian and American, have noticed that the absence of this signal made it difficult to get to sleep easily. Many of them hook an arm round something, or wedge themselves into a corner – anything so long as the body gets the pressure signal. This first night you'll probably use drugs to help.

As you doze off, the Skylab cluster will have its own set of noises to get used to, just like any house. Mostly you'll hear small valves opening and closing in the air systems. There may be the occasional short pulse from a tiny directional jet down at the other end of the cluster, as the computer corrects the alignment of the Telescope with the sun. So the night goes by with small poppings and the odd muffled tap. Consider yourself lucky that's all it is. During the final weeks of training the last Apollo Moonflight crews went to sleep with tape-recordings of the noises their Lunar Module would be making on the Moon. When they're down on a landing site they say there's so much popping and banging of valves and creaking of the structure as it reacts to the heat of the sun, that you can hardly hear yourself talk. And if Tom Stafford, Commander of Apollo 10, is anything to go by, they've only just got over the noise it makes getting down there – he says it's like flying a bathtub with somebody banging on the outside! Compared with their discomfort in hammocks at one-sixth gravity in a cabin where there was hardly room to scratch yourself at night, you're in a luxury hotel.

Next morning, after a wash, you zip on a set of cloth coveralls – the standard dress in Skylab unless you're going out for a walk. Somebody will have to, some time: there are experiments set on the outside of the cluster – paint and material specimens being exposed to the sun's light and to cosmic radiation – that have to be retrieved. The walk starts from the Airlock Module, where there's a small hatch opening directly into space. After first making sure that the hatches at both ends of the Module are closed sealing off the area from the rest of the structure, the crewman drops the pressure, and when it's all the way down to zero, he opens the hatch. After a while on Skylab, having somebody outside will become routine. In fact, short of floating all the way up to the Airlock Module, with its four windows, it will be the only way to find out whether the cluster is in darkness or sunshine, since the laboratory and wardroom have no windows themselves.

During the time you are on board Skylab the most important object of experiment and examination will be *you*. Well before the death of the three Russian cosmonauts on their return from the Salyut station in 1971, which aroused so much speculation before the final report was published, it had been known that zero gravity affected the human body. There is a certain amount of cardiovascular deconditioning, as the muscles of the heart get lazy from the lack of effort needed to pump the blood around the body. A change takes place in the bones. After a few days there *can* be marked symptoms of fatigue following a work load that on Earth would be considered light. So much has yet to be learnt about the long-term effects of weightlessness that Skylab, in one sense, is a leap in the dark. Some time during the first year of its life, a crew will be sent up to stay there for nearly 3 months. It may be, however, that zero gravity affects individuals differently after much shorter periods than that, so as you carry out the first of the bio-medical tests on yourself, remember that perhaps 2 weeks from now the results of these tests may cut the flight short. If one of the crew has to go, all of you must.

The medical tests are the most stringent you will ever undergo. They have to be. On their results rest the design of future space stations. If it turns out that man cannot live in a weightless condition for any significant length of time, say over 3 months, then the next generation of space stations will have to have some form of artificial gravity. At present it looks as if the easiest way to induce that gravity will be to spin the station slowly. That causes everything inside, including the crew, to stick to the wall opposite the station 'hub' with a force exactly equal to that with which we stick to the ground on Earth. But this operation can only be carried out on a station built to spin. Skylab isn't. That is not because the Americans (or indeed the Russians, since Salyut is also non-spin) ever believed that zero gravity would not cause problems. Even back in

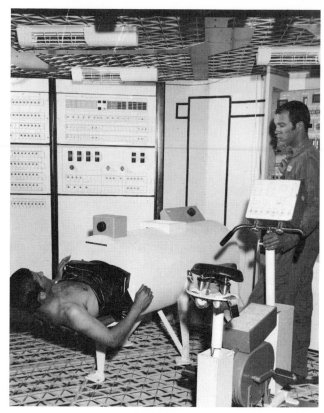

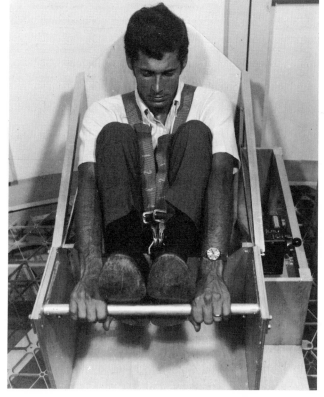

Left Getting your head down on Skylab – literally.

Top right The Lower Body Negative Pressure experiment.

the days of the all-out race for the Moon, Mars, and everything else in sight, the scientists knew that longer flights would need zero-gravity preparation for the crews, including considerable spells of practice weightlessness near enough to Earth to get back quickly if things went wrong. If flights to Mars and beyond ever do become feasible propositions, Skylab will have given invaluable assistance to the mission planners. But for the more mundane results expected from Skylab, from a purely experimental point of view somebody has to find out what effect 3 months and more without gravity does to a body. That is not to say Skylab was built non-spin in order to carry out these necessary tests. It was simply cheaper to build it the way it is: a rigged-out Saturn V third-stage fuel tank. Beggars can't be choosers – even when they're NASA.

There are no less than seventeen different checks on how your body reacts to being in Skylab. What happens to the mineral balance is checked by measuring the daily body weight. In zero gravity this can't be done with a weighing machine, so a complex box-like structure, not unlike a home rowing machine, is used. You fit yourself into this every morning, and tighten the structure around you. What is being measured is in fact your body *mass*, and from previous knowledge of your vital statistics the computers can work out mass change and relate it to changes in your weight. These data are related to the amount of food and liquid intake, and the amount of waste products excreted. The urine alone is subjected to ten different analyses, and the faeces to six. Blood analysis is also carried out. Other urine and blood tests determine what happens to your total body fluid, and in what ways weightlessness changes its volume and composition. Faeces are also checked simply to see how much is generated in the weightless state. The mass measurement is done by placing it in a sophisticated spring balance. When the pan at the end of the spring is pulled away and released, the oscillatory behaviour of the spring is affected by the material in the pan, and from this the faeces' weight can be determined.

Bottom right The orbiting weight-watcher using Body Mass Measurer.

But what about the mysterious heart deconditioning? Two experiments check the behaviour of your heart by placing it under stress. In one test you fit the lower part of your body and your legs into a barrel-like device with a seal that closes around your waist. The air is then pumped out of the barrel, subjecting the legs and lower body to negative pressure. This simulates the downward pull on the blood that would occur if you were standing upright on Earth. The other test is conducted using a Vectorcardiogram. Eight electrodes are fixed over your body as you pedal a bicycle ergometer at a predetermined rate. Your heart's reaction to this exercise, including pumping ability, electrical behaviour and general state, is monitored throughout. The observations are also made before and after the exercise.

A number of blood reaction tests are also carried out. The effect of solar radiation is examined in the blood cells, as well as their ability to react to drugs. The cells are monitored for any changes in their number, size, behaviour and life-span. But perhaps the most unpleasant test of all is one which will try to make you space-sick. You'll be strapped to a chair, with goggles over your eyes, and then the chair will be spun at anything up to thirty times a minute. Several astronauts and cosmonauts with no previous history of motion sickness have felt nausea in zero gravity – exactly why, no one yet knows. But if men less well trained or in worse physical condition than very healthy spacemen are to go into orbit at all, this particular mystery will have to be solved.

Even when you stagger off into mid-air after the first time through the tests, and zip yourself into your hammock for a night's sleep to recover in preparation for doing it all again tomorrow, you'll get no peace. One of the mission requirements is for you to wear a special helmet that is wired to a recorder that monitors your brain activity while you sleep. And as you nod off, tiny accelerometers in the cap will measure and record how much your head moves!

If all this makes you feel that Dave Scott was right when he said spaceflight was all minutes and seconds filled with things to do, take a look at the checklist of other experiments the Skylab crew have to complete before they come down. This is from the Mission List:

Habitability/Crew Quarters	The major purpose of this experiment is to obtain design criteria information for crew quarters construction in advanced spacecraft/space stations.
Astronaut Manoeuvring Unit	This experiment tests two methods of astronaut propulsion outside the spacecraft. The tests are conducted inside the laboratory.
Manual Navigation Sightings	Extensive checks on the ability of the spacecraft navigator to evaluate his position using a handheld sextant.
Crew-Vehicle Disturbance	This experiment measures the effect of crew movement on the stability and movement of their spacecraft.
White Light Coronagraph	Measurement of the sun corona from 1·5 to 3 radii out from the surface.
Spectrographic X-ray Telescope	Measurement of spectra of solar flare emission.
Scanning Ultra-violet Polychromator Spectroheliometer	Photo-electrical recording of solar images in six spectral lines.
X-ray Telescope	Total solar X-ray flux measurements.
Zero-G Single Human Cells	Measurement of the effect on growth of isolated cells.
Circadian Rhythm – pocket mice vinegar fly	Experiment determines the stability of the 24-hr cycle in the body of the subjects.
In-flight Aerosol Analysis	Determination of aerosol-particle concentration and size distribution.
Radiation in Spacecraft	Measurement of the absorbed radiation-dose rate and total radiation dose inside the laboratory and docked Apollo.
Expandable Airlock Technology	A test of composite materials proposed for use in expandable structures under conditions of long-term space exposure.
Thermal Control Coatings	Test of the environmental effect of near-Earth space on selected thermal coating materials.
Zero-Gravity Flammability	Tests on the flammability of non-metallic materials in a spacecraft environment.
Materials-Processing in Space	Five individual checks on the feasibility of electron beam- and thermo-welding in zero gravity and examination of molten metal flow, freezing patterns, thermal timing and surface tension
Precision Optical Tracking	A laser radar system for accurate tracking of the Saturn vehicle during the early phases of launch.
Coronagraph and Contamination Measurements	The determination of whether or not an induced atmosphere envelope is present around the Apollo spacecraft in flight.
Telescope Contamination	Measurement of sky background brightness caused by solar illumination of contamination particles floating around a spacecraft – in this case, the telescope.

Nuclear Emulsion	A check on the physical and chemical characteristics of primary cosmic radiation hitting the Earth's atmosphere.
Ultra-violet Stellar Astronomy	Examination of early-type stars with special optics.
UV X-ray Solar Photography	Check to obtain information for the prediction of solar flares and storms.
UV Airglow Horizon Photography	Photography of the twilight airglow and Earth's upper atmosphere in the visible and ultra-violet wavelengths.
Gegenschein/Zodiacal Light	Measurement of the surface brightness of the night sky.
Particle Collection	Collection of small micrometeoroids at satellite level.
Galactic X-ray Mapping	Mapping of galactic sources of X-rays of very low flux.
Ultra-violet Panorama	Photometric data collection of stars and dim sources in the middle and near ultra-violet.
Multispectral Photographic Facility	A six-camera multispectral photographic analysis of the Earth's surface in terms of crops, forestry, underground water, snow cover, water temperature, etc.
Infra-red Spectrometer	An assessment of this type of Earth-scanning corellated with ground-based observations.
Multispectral Scanner	Measurement of radiation reflected or emitted from selected sites.
Microwave Scatterometer, Altimeter, and Radiometer	The application of radar and passive microwave systems to study the Earth from space.
L-band Radiometer	Measurement of the brightness temperature of the Earth's surface along the Skylab track.

There's no doubt that Scott was right. In every one of the 28 days you stay in Skylab, there will be little if any time for leisure. It costs too much to send you up there just to enjoy the view. But it is the view that every astronaut talks about when he gets back. In Skylab you'll have to float out to the Airlock Module, where there are four special windows designed for optical observation of the Earth's surface. Any astronaut you talk to will tell you that the sight of the Earth from orbit is almost hypnotic. There's a rumour around Houston even today, 10 years after the event, that one of the early Mercury astronauts was banned from space after his first trip because he refused to come down, saying he wanted just one more look. When he did come down he'd used so much of his directional-thruster fuel getting good views that his re-entry was almost an emergency. As you look out from the window, the Earth below shows in incredible detail. It's claimed you can even see trains moving along the tracks in plains country, if you look hard enough. But should there be no trains about, it's almost certain that you would look down on a planet devoid of any sign of human habitation. From over 100 miles up the naked eye can see no mark of our existence, the planet might be empty of life. The predominant colour by far is a deep shimmering blue – the blue of Earth's oceans, that once caused a returning astronaut to exclaim that it looked like a blue oasis in the blackness of space. Along the immense edge of the curving horizon the atmosphere trembles, a light blue down near the surface, deepening with altitude until it darkens at the upper limit to black, as the air thins and the sun's rays have less and less to scatter and reflect them.

But put the eyepiece of a telescope or a telescopic camera to your eye, and the surface of the Earth leaps up at you, passing silently through your viewfinder at thousands of miles an hour. From up here, of course, the view of a continent takes perhaps 20 minutes to disappear from sight. You can see cities now, great brown and grey clusters, spilling out into the brown-green countryside around them, or huddled against a mountain range. Suddenly, as the straits between India and Ceylon come into sight, you understand the saying, 'The higher you are the lower you see'. Between the island and the subcontinent the sea is visibly shallower, a very much lighter shade of blue. This is one of the great prizes brought back from man's climb to orbit: the almost X-ray vision it gives. On the grand scale, you can see formations of rock plunge into the sides of mountains to emerge on the other side. Faults in the surface lead to oil wells at their end. From orbit you see that the faults fit a pattern, and that many of the features where you know oil deposits have been found are repeated in areas as yet unexplored. You can see the passage of the ocean currents at great depth, shifting and streaming the silt along round undersea hills and valleys, carrying with it plankton-rich layers that will be filled with browsing fish. There is virtually no end to the richness of the view you have of the turning planet. And it is in this ability to photograph and examine the surface of the globe, inch by inch, that the real value of man's so-called conquest of space lies.

The secret of its exploitation lies in the electromagnetic energy given off, or reflected, by every object in existence. Our eyes, miraculously efficient devices though they may be, see only a fraction of that energy – those radiations in the visible waveband. Our eyes sense remotely the existence of, and permit us to identify, objects, because they emit or reflect light waves, which are a form of electromagnetic energy. We can identify or interpret what we see remotely because our eyes construct an image of the object, and its shape and texture tell us what it is. The colour or spectral content of the radiation coming from the object also helps to identify it. This is what is happening as you lean against the window to gaze out over half a planet. But what you *see* is a tiny amount of what is there. There is a torrent of information pouring out from everything below. But it is coming on wavelengths our eyes cannot see. Each rock, for example, has its own particular set of radiations, that identify it as clearly as fingerprints. That set of radiations is called the spectral signature of the material. Set up an instrument to look for that one 'signature', and as your instrument scans the planet, it will reveal the material you are looking for, if it is on the surface. But if you know what to look for on the surface as a clue to the presence of your material lying beneath it, then you simply set your scanner to look for the tell-tale surface object. Find whatever is associated with gold, uranium, iron, manganese, and you are one step nearer finding the ores themselves.

In the infra-red waveband, heat emission is instantly detectable. This again is an electromagnetic radiation of energy, and all objects give off that particular radiation to a greater or lesser extent. A healthy plant will give off a certain amount. When it sickens, the thermal emission level changes, even from the first day of infection, weeks before any change can be detected on earth. That doesn't mean it is impossible on the surface, merely time-consuming to check every ear in a thousand miles of wheatfield, for instance.

From orbit your instruments receive and transmit to the listening computers on Earth millions of bits of information on the scene that passes below: how much water there is, when a storm will hit a fishing fleet, where a potential rift in the Earth will deny safe urban development, how much snow on the mountains will melt and flood the valleys and towns below – all these and hundreds more. This is where the future of manned spaceflight lies, here where you float tethered to a nearby handrail, or standing with your shoes with the curious triangle that projects from the soles locked into that crisscross mesh you wondered about on your first day – here on board Skylab and the space stations that will follow. Within 15 years Skylab will be a curiosity, floating along with the thousands of other obsolete bits of metal that are today already watched by the traffic control computer at Omaha, Nebraska, at Defence Headquarters. Then more than now will the accuracy of that control be vital to direct the computers on board the shuttlecraft lifting up through the orbiting junkyard of near space to dock with the giant stations ablaze with light, manned by hundreds of scientists, engineers, doctors, agronomists, geologists, anyone of the thousands of specialists who can read the surface of the planet, and bring what it can tell them back down to the rest of us.

2   Deep Secrets

In the last quarter-century there have been fifty-five wars. Over a million people have died in them and various portions of the earth's surface have changed hands. Every one of these conflicts has illustrated the tired maxim that war is an extension of economic policy. With each victory one group of people have taken over power from another. The ultimate source of that power, whether it be military or ideological, is, and always has been, the vanquished country's natural resources. But unless present-day predictions are overwhelmingly wrong all those resources – be they oil, uranium, gold, copper, or the contents of the upper layers of the earth's crust – are likely to be dangerously low or nearly exhausted by the end of the present century.

Throughout recorded history mankind has fought and pillaged and destroyed for what little there was beneath his feet. Strangely, no war has yet been fought for an infinitely greater source of riches which lies, virtually untouched, across two-thirds of the surface of the earth. It is a vast territory that belongs to no one; it is, in international legal terms, a jungle, the property of the first and most powerful invader. The war that no one has yet fought is the one to gain possession of the oceans of the world. The day when that war finally breaks out may not be too far distant.

Already the signs are ominous. Nations are already making unilateral decisions about their fishing and mineral rights to the seas that lie along their coastline. The areas they claim stretch variously to 3, 12, or 200 miles from the shore. America claims a 3-mile limit and the right to exclusive exploitation of her continental shelf. Some South American countries claim a 200-mile limit to make up for the fact that their coastline rises almost vertically up from the ocean bed providing no continental shelf at all. There is a Geneva Convention defining the continental shelf around a country as going to a depth of 200 metres and 'beyond that to where the depth admits to the exploitation of the natural resources'. But that definition, as we showed in our last book, is now hopelessly out of date. Technology has given the advanced industrial countries the capability of exploring and exploiting the seabed to well below that depth, and such is her economic strength that the United States has a virtual monopoly among those nations with the equipment to carry out undersea exploration. America has already undertaken ocean floor exploration to a depth of 15,000 ft, and is building equipment that will actually enable her to work the seabed at that depth before the end of the 1970s. A recent American Capitol Commission on Marine Science Engineering and Resources suggested that in the light of the undersea free-for-all ahead, some form of international rule of law was necessary. The Commission suggested the following formula:

Despite the new electronic underwater eyes and acoustic scanners that can peer into the depths of the ocean, when there is a tricky job to be done frogmen still go down to size up the situation first hand. But there are limits to the pressures a human body can stand. (See p. 70.)

(1)   All countries should renounce claims to seabed resources beyond the 200-metre depth.
(2)   From the 200-metre point to the point where deep ocean begins, each country could, under international ruling, allocate licences for exploitation and receive revenue from it.
(3)   An international control should be set up to deal with the resources beyond that limit.

Obviously the American desire to control activity on a local level, while leaving the ocean bed open to all-comers, springs from the fact that she is best

c

equipped to exploit such depths: few other nations have either the technology to do so, or the means of policing deep water even if they had jurisdiction over it. It is therefore in America's best interests to offer smaller nations something to get on with while she herself moves into deep water where the richest prizes lie.

As for Russia, there seems to be little or no desire on her part even to discuss the situation. With her own considerable land resources yet to develop, there would be no reason for her to agree to terms she might regret at some future time. For the moment seabed resources are not as vital to her economy as they are to America's, and it is therefore in her interests to keep the present free-for-all going until she chooses to join the scramble for the ocean's wealth. So at the moment the Super Powers show little sign of giving the protection of the law to whatever rights the 'have nots' might justifiably claim. It is this climate of potential piracy that has given rise to the near-explosive international situation that exists today. It doesn't make front-page headlines because as yet the man in the street has not been told just how big the stakes are. Yet the longer things remain unchanged, the greater the chance that commercial interests over the next few years will have a free hand to loot and destroy the last great untouched source of the world's mineral wealth. And the amount of wealth that lies waiting to be taken will attract the kind of exploitation that could, as the eminent British scientist Lord Ritchie Calder says, make the Klondike gold rush look like playtime in a kindergarten.

Fifteen years ago the oil production from off-shore wells was very small. Today ocean rigs provide no less than 16% of the world's petroleum production. Work both in academic and industrial laboratories has shown unseen mineral deposits to be attractive from both the technical and economic point of view. At the same time the increase in world population has made the exploitation of land resources more difficult. There is also some doubt that the continents can supply enough to maintain the developed countries at their presently accepted level of consumption of reserves. Furthermore, if the continent of Africa were to install a telephone system equal to that in the United States, it would require the total amount of copper known to exist in continental deposits throughout the world.

As exploiting the land becomes more costly it now looks as if the seas could provide a source of material that is not only plentiful but cheap. An international committee has reported that copper could be produced from the sea at about half the land extraction cost. The same is true for cobalt and nickel. It is also claimed that the investment risk is, unexpectedly, low. On the land, the major mining companies spend, in aggregate, about 30 million dollars in exploration costs for each major metal deposit they find. In the Pacific Ocean it would take a most clever exploration geologist to find a place outside the trenches and off the continental shelves and slopes where manganese-nickel-copper-cobalt deposits *cannot* be found. As for the design and construction of a mining equipment system to handle this wealth, the cost is estimated at less than a single Apollo Lunar Module – 5 million dollars.

Each year the world's rivers discharge thousands of millions of tons of material into the sea. Over long periods of time these materials are separated and refined, and end up in one of five places: held in suspension in the sea water itself, on marine beaches, continental shelves, the subsea floor, or the deep-sea floor.

Onshore beaches have been mined for some time. Because of the industrial action of the waves, which crush, grind and separate, what lies on the beach is almost the finished product: diamonds, gold, platinum, magnetite – to name but a few minerals already being found in quantity. But the offshore beaches still await exploration. It is expected that the magnetic and sonar devices now being developed will discover even more material there than has already been found above the tidelines, because they will be searching the ancient Ice Age beaches, themselves once above the tideline before the world sea level rose millions of years ago. The continental shelves cover about 10 million square miles, or about a fifth of the area of the land mass of this planet, and they are likely to contain at least as much wealth as onshore rocks, enhanced by the mineral content and active nature of the water that covers them. Potentially one of the richest areas so far examined is the Atlantis II Deep on the floor of the Red Sea, which sits within the local continental shelf. The area consists of a deposit of gel-like sediment lying about 2100 metres deep. The nature of the sediment promises relative ease of extraction. The layer itself is about 90 metres deep; core samples from only the first 10 metres indicate that about 50 million tons of salt-free solids, worth about £1000 million, are lying there ready to be taken. The subsea floor rocks off Japan, Canada, Finland and Great Britain already show promise. And these are the areas where exploration has already begun. This year, around Britain, oil strikes have been made that it is estimated will satisfy half the total oil demands of the country for the foreseeable future. The political implications are, to put it mildly, far-reaching.

Sea water, it is generally agreed, contains within it all the natural elements that exist. It covers about 140 million square miles of the planet's surface, at an average depth of 2·5 miles – in all about 350 million cubic miles, containing on average about 3·5% of elements in solution. That calculation is the interesting one: as a cubic mile of sea water weighs about 4·7 thousand million tons, the materials in it must weigh about 166 million tons. When the researchers perfect a cheap method of extraction, one of the places that will receive immediate attention is the Red Sea. It contains what are called 'orebodies' – areas of water so dense with dissolved minerals that they can be compared to lodes in an underground mine. No less than 50,000 times the normal amount of water-borne minerals are to be found in them.

But it is the deep ocean floor that will be the greatest prize of all. There, the natural actions of the sea work on a giant scale, separating and concentrating the mineral wealth. The most spectacular of all deep-ocean deposits, because of their vast extent, high grade, and what one mining expert calls 'their potential for restructuring the mining industry of the world', are the manganese nodules. These black-brown potato-size lumps lie scattered in huge fields all over the floor of the three major oceans. Each nodule is formed in layers, rather like an onion, by slow accretion, growing approximately 10 mm in every 1000 years. They are, by any standards, very high-grade ore. Most important of all, early analysis shows that if we were capable of mining 10% of the total deposit per year (assuming, in the worst instance, that 90% were inaccessible) the sea action would be replacing the nodules three times as fast as they could be removed.

These early discoveries indicate a potentially vast source of mineral wealth. But there are scientists who believe that we have not even begun to appreciate its full extent. And when the land resources of the super powers begin to run out after the end of the century, they will have to extend their boundaries across the ocean floor, in order to benefit from the mineral resources they will so desperately need. As they do so, they will have to develop the means not only of staking their claim, but of maintaining and defending whatever territory they occupy. Once those means are developed, and the equipment is ready to move, scientists and researchers will be involved in what is essentially a para-military situation. The time for discussing international agreements and safeguards will be past.

Already the tools are being developed. For the past 2 years an American ship, the *Glomar Challenger*, from the Scripps Institution of Oceanography in San Diego, has been criss-crossing the major oceans, examining the sea bed as part of a 4-year Deep Sea Drilling Project. So far the researchers on board have been

Left Unloading wire dredge basket filled with manganese nodule samples recovered after television survey. Site location is recorded and nodule samples are analysed aboard ship to determine metal content. Additional samples are sent to Deepsea headquarters for further analysis.

Inset photograph shows actual nodule deposit as received on *Prospector* television monitor. TV tapes are made of such deposits for study by scientists at Deepsea headquarters. Television camera mounted on tripod is transmitting this picture from approximately 18,000 feet below the ocean's surface.

Right Television camera and tripod prior to launching over the side of R/V *Prospector*. Fin mounted on tripod (*top left*) maintains stability as camera and tripod are towed at two knots a few feet off the ocean bottom. High-intensity lights are mounted on tripod for illumination.

finding out how and where the earth's crust is still moving and throwing up new mineral deposits, how the oceans are changing and how the continents are drifting apart. At the end of the project *Challenger* will have made the most comprehensive survey of its kind ever undertaken. The data accumulated will be comparable to that brought back in the past few years from near-space. The photographs and measurements taken by the astronauts in orbit gave geologists the kind of information they needed to pinpoint new sources of minerals on land. When *Challenger*'s secrets are revealed, we can expect the same kind of explosion in undersea knowledge.

In the meantime the search goes on. One of the latest 'aids' is a nuclear probe that can detect minerals in minute quantities – on the sea floor. Developed by scientists at Battelle-Northwest, at Richland, Washington, the probe will be used for geophysical mapping of the sea bed as well as for commercial exploration. Although the detection of certain minerals has been possible before, this is the first time an exact and minute analysis can be carried out on the spot. Within 3 to 5 minutes the probe reveals the kinds and quantities of many elements present

in a mineral deposit down to as little as a few ounces per ton. The probe consists of an irradiator and a detector. The irradiator sends out a low flux of neutrons from a Californium 252 source (a piece of the element weighing about 1/100,000th of an ounce). The radiation is absorbed by the minerals being measured. Then most elements react by producing gamma rays and the measurement of the gamma rays given off and their comparison with a computer memory store tells the operator exactly how much material is present, and its quality. To be able to conduct exploration, mineral strike and assay in a single operation will allow prospecting organisations to make money-saving decisions on the spot. A company using the nuclear probe could, for the first time in mining history, direct its engineers and equipment to a source of minerals, having already worked out to the penny the amount of profit it was going to make.

For excavation of deposits found close inshore, Ocean Science and Engineering has just developed the world's first submarine dredge – the Crawlcutter. Its body section is made up of two cylindrical modules at right angles to each other and connected by a crawl-through tube. The rear module houses a 750-hp electric motor that powers the dredge pump, as well as the electric-hydraulic system that powers everything else. The smaller 6-ft-diameter forward cylinder is the operator's compartment. He has three sets of instruments on the control panel mounted in the cylinder. One shows depth, attitude and orientation of the vehicle, another the motion of the dredge relative to the bottom, and the third shows the density of the material being pumped out and the pump pressures themselves. As the operator moves the dredge forward, a rotating cutting head eats into the sea bed, dumps it into a 12-inch pipe through which it is sucked over half a mile to be spewed out on the beach. Crawlcutter has caterpillar tracks of sufficient power to knock down and ride through a 6-ft-high wall of coral. To make sure the vehicle remains on an even keel, four built-in jacks can each lower a 4-ft-square 'pad' and raise the entire machine by 3 ft. Sticking up from the central structure is what looks like a chimney or snorkel. In fact, since the operator breathes air pumped in from the beach, it is neither, but allows the operator to get in and out even in the roughest swell. The ability to dredge the inshore beaches in any weather is the principal advantage of Crawlcutter, and it can also work at considerably greater depths than conventional surface dredges. It would be nice to think that this kind of development will encourage local authorities to maintain coastline amenities by repairing and restructuring beaches where necessary. What is more likely to occur, however, is exactly the opposite. Twenty years from now you may slide into the sea from your beach which has been built up to 8 ft, into waters crawling with machines looking for mineral deposits, and dumping what they don't want on the spot you've just left.

For deeper water, a French company, ACB, has developed a system for selective deep-water dredging. It consists of an underwater vehicle, remotely controlled via an umbilical cord from a surface vessel. The underwater dredge will also be connected to a float directly above it; when it needs repair or maintenance, the float merely hauls it up. The dredge is capable of pumping to the surface 500 tons of solid material an hour. Although its positioning is blind, since at these depths there is no point in installing television, this type of excavation would be carried out only after extensive exploration and mapping had identified the ore-bearing rocks. At that point three directional loudspeakers would be lowered into the water around that area. Signals from them would trigger off controls to guide the dredge back and forward in random fashion until the entire area encompassed by the loudspeakers was fully excavated. Then the loudspeakers would be repositioned for work to continue elsewhere.

The greatest amount of exploration and mapping in recent years has been in

The French deep-water selective dredger. The float on the right supports the dredging head. The whole operation is controlled from the ship.

the search for off-shore oil. Almost all the wells have so far been sunk in shallow waters, using oil rigs that either pump the crude oil immediately to shore, or into waiting tankers. The difficulties of an undersea oil operation has kept commercial interests from working in deeper water. But now two French oil companies have built and tested a deep-sea oil terminal. The strategic implications, as well as the cost-efficiency of keeping the crude oil underwater and close to where it came from, are very attractive. The French tanker consists of a number of storage tanks positioned around the well-head on the sea bed itself. Each tank is a vertical cylinder with a capacity of up to 600,000 barrels. Reinforced concrete sections and internal partitioning give it its strength. After construction ashore, the tanker is towed to its ocean site, flooded and sunk. As it descends to the bottom a system of floats, each of controllable buoyancy, is used to guide the tank with great accuracy to its final position on the sea bed. Once there it is connected by pipes to a central supply tower. The tower under test at the moment in the Bay of Biscay is 420 ft high, 23 ft in diameter and rises about 100 ft above the surface. It is known as an Articulated Platform, because the bottom is attached to a sea-bed platform by a giant universal joint, thus allowing the entire power structure to oscillate with the waves instead of resisting them, thereby avoiding considerable stress on the tower from the movement of the sea.

Early tests with models revealed one interesting fact. It has been assumed that a regular sea swell would impose a symmetrical wavy stress on the column. But it turned out that the speed of water movement in a wave was higher at the top and that this part of the wave produced the area of greatest stress. Because of this, the ideal structure was found to be one that would lean in the direction of faster water movement, hence the universal joint. The present Biscay test structure houses six men, as well as the pumping equipment necessary to bring the oil from the tanks to the surface vessels. There would appear to be no limit, other than cost, to the size of the tower, and therefore to where it could be placed for operation. The successful testing of the tower in the Bay of Biscay will almost certainly mean an increase in deep-sea drilling for oil, and the consequent increased need to guard against spillage on a scale which would make the *Torrey Canyon*'s contribution look like a puddle.

It may look just like any other electric bulb. But this one is different. It is encased in an extra glass envelope.

Looking ahead to the need for underwater construction and repair on a production-line scale, technologists are now examining the particular problems created by working underwater. Simple activities like welding are no longer so simple. The main problem is that, due to the surrounding water, any weld carried out by a diver cools much more quickly than it would in the atmosphere, and can become dangerously brittle. So an American company is now working with the first diving and welding equipment that allows divers to carry out their repair work in a dry environment, called, because they inhabit the work area, a 'habitat'. The new technique was first worked out when a gas line broke in the Gulf of Mexico. At that time no equipment existed that would repair the pipe, replace the broken section, and weld the pieces together underwater. The new system that would do it all has now recently been completed and tested. It consists of a cradle of pipes and girders that hold a metal hut, the welding 'habitat' where the divers work. A guide-rail ensures that the habitat can be moved 4 ft horizontally or 8 ft vertically for positioning over the broken section of pipe. Weighing about 165 tons in air, the habitat is lighter by 80 tons when submerged, and when the water is pumped out, the working weight drops to 20 tons. Beyond each end of the habitat is a giant, hydraulically-operated clamp which grips pipes up to 48 inches diameter and is capable of the same horizontal and vertical movement as the habitat. It is also capable of swinging 17° laterally and vertically, to allow for accurate pipe connecting. The framework holding all this measures $167 \times 32 \times 27$ ft and carries the rest of the equipment necessary. This includes a life-support system via an umbilical cord to the surface vessel. Constant monitoring maintains oxygen and carbon dioxide levels within safe limits, with a provision for automatic injection of oxygen at any time. The crew enter the habitat either through a side tunnel or, if the local sea bed is not too muddy, under the walls. Two TV cameras inside the habitat ensure visual contact with the surface, as the crew move the pipes together with two 10-ton hydraulic cranes mounted in the roof. In order to ensure that the welders are acclimatised to the conditions at working level, pressure breathing is carried out in the diving bell which lowers them to the repair area, and once there the test crew have shown that they can work up to $10\frac{1}{2}$ hours at a stretch. With systems like this, another major limitation on building structures like underwater oil terminals is removed.

The lighting systems needed for working at depth have always been cumbersome and expensive. If a diver's lamp burned out he had to return to the surface to replace the bulb, since up to now the pressures at deep level have meant enclosing lights in heavy protective casings. Now a British firm has produced a light that solves the problem. It consists of a very bright quartz-halogen bulb, with an extra glass envelope around it. The envelope is specially hardened to withstand pressure and sudden changes of temperature without cracking. If it should burn out in deep water, all the diver has to do is unplug and replace. And it's ten times cheaper than its predecessors.

This increased availability of light now means that less expensive and sophisticated television equipment can be used. TV has for some years been essential in underwater work both from an operating and exploration point of view. You need to be able to study the area indicated by early surveys in order to pinpoint mining or salvage work, and, as with the repair habitat, visual communication is essential for safety reasons. In a recent experiment with the British submersible, *Pisces*, television pictures were successfully broadcast from the site on which the sub was working. Once the crew decide to broadcast, a sonobuoy is released and floats to the surface. It carries with it a cable, bringing the video signal from the camera hundreds of feet below, and a transmitter, to beam the signal to a receiver on shore. The test proved that the day is not far off when a survey submarine will be able to transmit the findings of its own minerals probe, show geologists on land what the area looks like, and begin the first claim-jumping without ever surfacing. If techniques like these fall into the wrong hands, the consequent need to police areas that might be under theoretical control by a country which was unable, for reasons of cost, to carry out its own surveys could be virtually unanswerable. With nothing but an undetectable tiny buoy carrying a remote transmitter, any pirate underwater speculator could work uninterrupted – the underwater 'claim-jumper' of the 1990s.

But if you're working on a small budget and there's no one in sight, there's a much cheaper way of looking around the ocean floor. From studies by the Centre for Advanced Marine Studies in Marseilles a French company has constructed a very simple sledge-like vehicle called the Troika. It needs no motive power of its own, since it is drawn by a surface vessel. It carries an acoustic transmitter to indicate its position, and an automatic film camera with a flash. Pictures are taken as the sledge travels along the bottom on its streamlined 'runners'. Tests so far have shown that Troika can provide pictures of the ocean floor down to the hitherto unknown depths of 30,000 ft, a limitation imposed only by the structure of the lights on board.

It may be that early surveys of the sea bed will indicate the desirability of setting up some form of undersea living quarters. These could take the form of a base camp for extensive mining operations. They could equally well be military submarine fleet headquarters. The deep-ocean bed offers attractive opportunities for concealment, and American military circles are known to favour the development of ocean floor complexes. For these reasons, engineers have been looking at new and exotic materials, like the latest alloys, searching for the ideal undersea construction material. And what they have discovered is that the best material of all may turn out to be concrete.

Already tests have shown that giant 50 ft to 100 ft concrete spheres and cylinders, which would permit work in a shirt-sleeve environment, are feasible down to depths of 3000 ft. The potential applications for large concrete structures are numerous: containment structures to house nuclear reactors located off the coasts of major cities; garages for the supply, maintenance and repair of commercial submarines; military defence stations to monitor underwater activity and provide logistic support to submarines that need never surface; mineral refining plants; oil-drilling enclosures; and even tunnels extending from the coastline to deep water. Existing tunnels, like those under the Thames and the Chesapeake Bay Bridge Tunnel in Virginia, have shown that concrete structures will stand pressure down to 100 ft. Now concrete spheres have been subjected to pressures found at 27,000 ft by the US Navy Civil Engineering Laboratory. Even the introduction of windows and hatches into the side of the spheres failed to alter the implosion depth, so long as the materials used were of the same elastic properties as the hull material itself. It was found, for example, that titanium or steel used at these points significantly *reduced* the implosion depth, unlike a more compliant material like fibre-glass. At the same time research has produced a substance which will make concrete totally impermeable to water, resistant to its corrosive effects and twice as strong. This is achieved by injecting a form of liquid plastic into the tiny cavities that are present in the hardened concrete, and then solidifying the plastic. The crews living and working in the concrete structures will still have to put up with limited views of the undersea area around them, since at the moment the area of wall that can be given over to windows is necessarily small.

The problem of visibility underwater is the subject of other research at the US Naval Laboratories. One early product of that work has been the development of NEMO (Naval Experimental Manned Observatory). This submersible is quite unlike any other. The main aim of its designers is to give the undersea observer

Concrete spheres like this one are proving that concrete may be the best material for deep-water structure. It is tougher — and cheaper than everything else.

The first oceanographic auto-buoy to work on wavepower. A hollow tube extends vertically downward from the bottom of the buoy. As the buoy moves up and down with wave movement, air trapped at the top of the tube is forced upward, driving a turbine blade in a mini-generator to produce electricity. As the wave (and water level) subsides, valves open to permit more air to enter. They then close, and the cycle begins again. The wavepower generator has been developed by the Japan Defence Agency.

all-round vision. Most research subs afford a very limited view out of tiny port-holes set into the hull. NEMO provides total all-round visibility. It is basically an acrylic plastic bubble, which is a very attractive material for use down to 1000 ft. As well as being transparent, it has very high compression strength. It can also provide a lighter pressure hull than steel which is usually over-designed to avoid having thin sections that might buckle. But best of all, the physical properties of acrylic plastic are actually enhanced by lower tempera-tures. NEMO has undergone tests in the Bahamas to 500 ft. The craft manoeuvres using two tiny reversible motors mounted on the platform below the control bubble. These can propel NEMO slowly forwards and backwards or rotate her slowly. NEMO's power, like that of so many submersibles, comes from a battery. Electric power is the ideal source of propulsion for many reasons: no need for massive fuel tanks, ease of use within the submersible and relative lack of pollution of the environment outside. One of the drawbacks, however, has been the weight factor. In the very big subs generators are needed, and in the smaller research craft conventional batteries force design constraints on the builders. The development of a new fuel cell by the French may change all that. The Alsthom organisation has produced a unit that will provide no less than one kilowatt of power for every cubic decimetre of fuel cell structure. Each cell is divided down the middle by a semi-permeable membrane of microporous plastic about 100 microns thick. The outer walls of the cell are formed by two corrugated electrodes, coated on the inner side with a catalyst. Into the space between the electrodes and the central membrane, liquid reactants are injected. Most of the research work has been carried out using hydrazine and hydrogen peroxide, although other mixtures are being considered, like methanol and air. As the reactants flow through the space between the electrodes and the membrane, they seep through the tiny holes in the membrane, and react to produce full power within half a second of activation: a positive charge through one electrode, negative through the other. The total thickness of each cell, electrode to electrode, is half a centimetre. In practice a number of cells are packed alongside each other to form modules, which in turn form part of a modular stack. The greater the number of cells in a stack, the higher the power. Each stack is clamped between endplates equipped with fluid feed and discharge tubes which connect with the feed and discharge passage running through from cell to cell. These tubes allow stacks to be easily 'plugged' into each other. Perhaps the most important factor in operating the fuel cell underwater is the reactants themselves. Being of a density only slightly higher than that of water, they can be carried in flexible plastic tanks capable of going to virtually any depth. As the reactants are used up they produce small amounts of nitrogen and oxygen, which are safely discharged into the sea. The liquid by-products also produced can be collected in another plastic tank. This avoids pollution and, perhaps more important to the commercial operator, maintains the negative buoyancy of the craft at almost the same level, since these by-products weigh very little less than the reactants.

The range of possible uses for the system seems considerable, whether for diving saucers, small interceptor submarines, underwater habitats or subsea oil wells. Yves Cousteau's diving saucer, for instance, carried conventional storage batteries weighing about 380 lb. The saucer recently used the fuel cell, which produced all the power it needed from a cell stack a quarter the weight of its conventional batteries. The fuel cell is the next logical step in the development of undersea power systems, just as it has been in the move from aero- to space-flight. And it could prove to be of great value to those developing countries whose budgets do not run to long-term submarine power systems operating on nuclear fuel.

With all the resources of a super power at her command, America is the only country in the world that can afford to use nuclear power in a non-military submarine, the NR-1, the world's first research submersible with a nuclear power plant. Launched 3 years ago, the NR-1 is the property of the US Navy and is shrouded in secrecy, but bits and pieces of information leaked so far reveal a boat with considerable capabilities. She costs somewhere in the region of 40 million pounds. Her basic task is to demonstrate the feasibility of nuclear propulsion in a small sub, able to perform a variety of oceanographic and military tasks. NR-1 looks like her larger sister combat vessels, with a cylindrical pressure hull. She is 140 ft long and has a 12-ft beam, submerged displacement is 400 tons, and there is room on board for a crew of seven. There are twin screws, and mobility is provided by four ducted thrusters, two in the bow and two in the stern. The sub also has diving planes mounted on the sail and a conventional rudder.

NR-1 is designed as a multi-purpose deep-dive vehicle working for the US Navy's Deep Submergence Systems Project. For oceanographic work, she has viewing ports, external lights and TV cameras, and a remote-controlled manipulator arm capable of picking up small objects. As for the kind of operations she is at present engaged on, the Navy has revealed that NR-1 should be able to map thermal and current patterns in the ocean; service manned or unmanned undersea installations; aid in the further development of deep-dive research; and provide the best means of carrying out Arctic under-ice work in preparation for the submarine oil tankers of the future.

As submerged traffic increases, engineers continue to work on new types of structures for use in the sea. Many of the designs spring from our new-found knowledge of how structures behave in the underwater world. The more we learn the more we realise the tremendous power exerted by moving bodies of water. As a result bridges are now being designed deliberately *not* to just stand and take the shock. Such a structure, the conventional bridge with solid foundations, would have to be prohibitively strong and expensive, if it were, say, to cross the Straits of Gibraltar. The Straits are 8 miles wide and about 2800 ft deep; currents move up to 7 mph and wind speeds of 60 mph have been experienced. An American company, Ocean Science and Engineering Inc., has designed a new kind of bridge to solve the problem. Whether it will be built or not remains to be seen. What is important is that the design shows that it is feasible to think of uniting Africa and Europe with a four-lane highway.

The basic task was to put a bridge across a gap too wide for a suspension bridge span and too deep for ordinary concrete piers. The new system is called the 'tension leg' bridge. It works rather like a super-taut mooring buoy. Three large cylinders form a structure held well down below their normal floating level by three vertical cables anchored to the sea floor. Each cable is composed of seven 6-inch-diameter cables bundled to form an 18-inch-diameter cable with a breaking strain of more than 12,000 tons. The three 45-ft-diameter cylinders offer a total of 51 million lb of buoyancy, four times that needed to support the bridge leading above them. The result of the pull by the cylinders on the cable is to produce no vertical movement in the cylinders and only very slight horizontal movement. There is, however, the problem of joining the individual sections of the bridge together. Each section can, on occasion, move independently of another. This can occur due to currents, winds, or the passage of very heavy loads across the bridge. The designers have added 'hinges' 200 ft long between the main sections; these hang on paired groups of short cable slings, supported by rockers, to allow the gaps between sections to open and close without changing their vertical relationship. Included in the design is the capability to take the '100-year wind', the name given to freak conditions that might exceed normal violent weather states by 40%, and that therefore might not be expected to occur more than once a century. Even on such a rare occasion, the bridge design means that the road surface along the top would tilt by no more than the normal amount of camber built into landbased roads of today.

The problem of handling that kind of wind strength has been avoided by an English group of engineers who have designed a system for taking a road across the Messina Straits between Italy and Sicily. The entire structure they have designed is submerged. It consists of three giant tubes, carrying two 3-lane highways and a 2-track railway, that would run 120 ft below the surface of the 2-mile-wide straits. The design was entered for a competition sponsored by the Italian government for a scheme to get the Autostrada del Sole, Italy's major motorway, across the Straits and into Sicily.

The construction of the tunnel would be relatively easy. Lengths of 300 or 400 ft, cast in concrete, would be sealed and floated into position. They would then be flooded enough to let them sink to tunnel level where they would be attached to the section already in place and pumped dry. The seal would then be broken between the new section and the rest of the tunnel. A computer analysis of the current behaviour in the straits revealed that the tunnel could be held in position by mooring ropes. Two ties fixed to either side of the tunnel would be attached to a common mooring cable anchored on the sea bed. This would prevent twisting in the current, and allow considerable give during minor earth tremors, which are common in the area. Above all the underwater 'bridge' would avoid the hazards of high wind and expansion or contraction due to atmosphere temperature changes. And it would be cheaper than building a conventional bridge.

The 'tension leg' bridge – used where it is too wide for a suspension bridge span and too deep for ordinary concrete piers. (See page 49.)

Anchoring objects on the sea floor is not a new problem; it has been around for as long as sailors have had to moor ships and buoys. But in the last few years, with the advent of the supertanker, the problem has taken on an entirely new dimension. As the tankers get bigger, the ports of call which they can visit grow fewer, as do the number of estuaries or offshore shallows deep enough to take the new giants. The only economically feasible answer to the problem is to moor the tankers out in deep water, and carry out all loading, unloading and maintenance operations well away from the coastline. But how do you anchor a buoy securely enough to take a fully loaded, 250,000-ton ship and hold her through any storm, as she swings in all directions in order to keep her bows into wind? One French company, Cocean, have answered the needs of the supertanker with a superbuoy. Several of them are already installed and operating successfully. They are carefully situated in waters that are either well off other shipping lanes or nowhere near areas of population likely to suffer or complain about loaded supertankers going through trials nearby.

The secret of holding back the tanker in the roughest weather is to put a giant plug in the sea bed and tie the ship to it. First, a diver goes down to the plug site with a hand-operated drilling rig. This is a hydraulically operated device, about 10 ft high, standing on four legs. Each of the legs can be telescoped to ensure that all four footpads rest evenly on the sea bed and that the drill goes in vertically. The drill itself is a complex affair: as the main head cuts vertically downwards, smaller drill heads mounted on the shaft and pointing outwards widen the hole. This drill-cluster operation produces a hole about 20 ft deep and 3 ft wide without the need for a single drill of the same dimensions, making it easy for one man to position and operate it.

A quarter-of-a-million-ton oil tanker will swing at anchor above this metal and concrete plug planted firmly in the sea bed. The other end of the chain connects to a giant buoy on the surface. This new technique has been developed for super-tankers too large to get into port.

When the hole is ready, a massive concrete plug is lowered to the diver and inserted. Once it is cemented in position, a chain is attached to the top, the other end of which is fastened to the underside of the new superbuoy. Because of the tremendous size and strength of the chain and its anchor, the buoy can if necessary be made big enough to house a control room and a helicopter pad. The designers are looking forward to the day when hoses leading to a shore refinery or storage complex can be fitted to the buoy, turning it into a kind of remote loading and offloading facility. With the tanker tied up to its own personal supply line in this way, the once hazardous job of handling oil in that quantity close to shore disappears. Above all, the operation can take place well away from shore. Unfortunately systems like these take away the incentive to look for some other way of transporting vast quantities of oil round the world, by literally giving the supertankers more room to manoeuvre. A few years ago it was thought that onshore handling problems would eventually halt the development of supertanker fleets. Now it appears that the technologists have removed that obstacle, and in doing so increased the probability of further pollution of the seas and coastlines.

One of the main restrictions on handling pollution of the ocean is that we don't know very much about how the ocean itself behaves. This is why, when oil slicks or giant pockets of industrial pollution float ashore, there is no means of knowing exactly where they came from. Without that kind of information it is impossible to stop the dumping of filth in the sea. But we are learning more and more every day about the currents that move in such apparently random patterns across the oceans. Until quite recently measuring tidal flows, subsea currents and the movements of vast bodies of water in deep ocean has been a rather hit-and-miss affair. Survey ships have had to lower current meters over the side, take a reading, and move on, and it has taken nearly 40 years to build up the meagre amount of current and flow information we have. Now a British organisation, the National Institute of Oceanography, has made a device which could revolutionise the entire information-gathering process. It consists of a number of lines of miniature battery-powered current meters. These include sensors capable of measuring the strength and direction of movement in the water, and tiny magnetic tape-recorders. The recorders can store about 6 months' information on one spool of tape. The meters are dropped over the side with an anchoring device at one end of the line and a large spherical buoy at the other. When the survey ship returns a few months later it sends out an acoustic underwater signal, which triggers off a small transmitter on the anchoring device, on which the ship homes. As it closes on the target area the ship sends out a prolonged signal which fires off an explosive bolt in the anchor, cutting the line so that the buoy, together with its meters and their spools of information, floats to the surface. Just in case the ship should still be too far away for a visual fix on the buoy, it is fitted with a radio transmitter to send out a signal as soon as it breaks surface. Analysis of the tapes will tell researchers a great deal about the behaviour of the sea in that area over a 6-month period and, by inference, something of the ocean dynamics at large during that time. With detailed information like that about the movement of surface water, shipping could begin to use the seas in a much less haphazard manner than they do at the moment, when every voyage is much like a plane flight where the pilot knows about the storms en route, but little or nothing about the winds.

The currents are the winds of the ocean, and can sometimes add several miles an hour to a ship's speed. That kind of information would be of particular benefit to the ocean-going cargo-shippers, for whom speed of delivery is vital. One of the ways in which cargo-carrying has been made more efficient in recent years is the development of containers. Slotted together on board specially designed ships, they permit a much more cost-effective use of space, and render loading and unloading much more quick and efficient. The same principle has now been applied

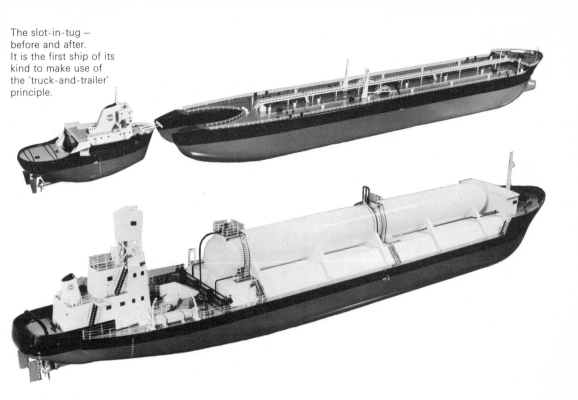

to the design of an entire ship. Built in the US by the Alabama Dry Dock and Shipbuilding Company at Mobile, Alabama, it's the first-ever sea-going equivalent of the truck and trailer, only in this case the 'truck' pushes. The engine and control sections form a small ship in themselves, acting like a tug, that slots into the aft end of a giant cargo-carrying barge, over 500 ft in length. The combination forms an ocean-going cargo ship until port is reached, when the tug separates. This allows the time-consuming business of unloading and reloading to take place without wasting the potential working time of the tug, which can pick up another fore-section and leave. The economic advantages are obvious. The system is also a lot less cumbersome and dangerous in high seas than the conventional towing methods used at present. The new vessel is made of Aramco steel, and is due for launching in 1972. Although this concept has been around for some time, it is the first time it has been applied on such a scale. It remains to be seen how well the system will react to the high seas. The ship will probably be given trial runs initially on coastal or sheltered waters before venturing out on longer trips.

As the cargo ships get bigger, the business of manoeuvring them becomes more and more complex, particularly in shallow coastal waters and when entering port. A number of collisions have already revealed how serious the problem is. The really big supertankers, for example, require several *miles* stopping distance, and a fast change of course in an emergency is out of the question. With this in mind, a French company has designed a system for steering the big ships which is based on the jet engine principle. It is claimed that this will enable a supertanker to turn in her own length. The secret lies in jetting water out at close to right angles from the bow of the ship by means of a turbine blade set into a small tunnel which opens, below the waterline, in the ship's bow. Aft of the blade the tunnel forks, leading to openings on either side of the hull a few feet back from the bows. These openings have hydraulically operated 'doors' which can be worked independently to close off either of the tunnel exit points. Under conditions where the steering machinery, the 'Y-jet', is not needed, water simply passes into the tunnel through the turbine blades, causing them to idle, and back out through the two hull exits. If a sharp manoeuvre is needed, one of the exit doors is closed and the turbine blade spun at high speed by an electric motor buried in the bows. The blade sucks in water and expels it at very high velocity through the remaining exit. As the water rushes out, almost at right angles to the hull, the force pushes the bows of the ship rapidly round in the opposite direction.

This system can also be used to slow the ship down. In this case both doors are left open when the turbine blade is switched on, and the water that is forced out on both sides breaks up the smooth flow of water along the sides of the ship and causes her to pass less easily through the sea, in the same way that spoilers break the airflow over an aircraft's wing, and cause loss of lift. The designers claim to have found this to be an ideal system for manoeuvring tankers of 250,000 tons deadweight at their most critical speeds, from 1 to 10 mph, the point at which this kind of ship is virtually out of control. The Y-jet will also work when the ship is going backwards at speeds up to 5 mph, because the turbine will still draw in water in significant amounts; and if the blade direction is reversed, then the system can be used to help propel the ship backwards. So far the Y-jet is still on the drawing boards but extensive model tests have been carried out which indicate that the system works perfectly. Hopefully it will soon be in use, as the number of supertankers grows, and their manoeuvrability becomes more and more a matter of concern, not just to their navigation officers, but to coastal regions. The potentially disastrous consequences of more frequent collisions between the 700 supertankers that will probably exist by the end of the decade make it all the more imperative that some efficient system to give greater manoeuvrability be adopted

The Y-jet under test.

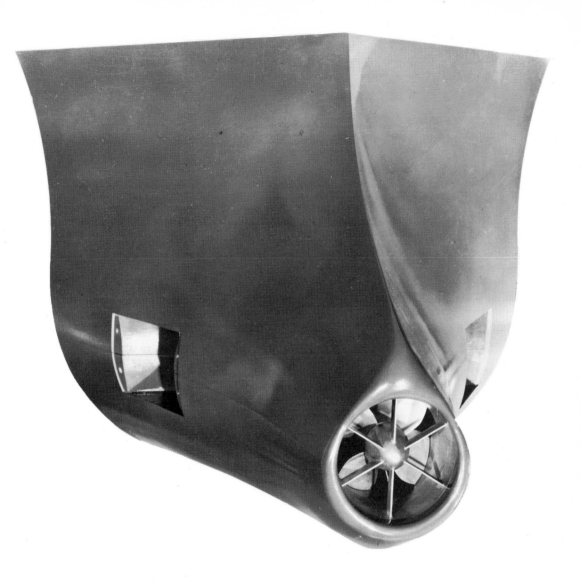

The Y-jet is designed to steer big ships using a principle based on the jet engine. By opening and closing doors at the base of the hull, big ships may now manoeuvre with startling speed and accuracy.

without delay. It is, on reflection, extraordinary that oil and chemical carriers should be built on the scale they are today with control systems that are virtually unchanged since the invention of the marine engine.

One surprising area where movement control is being developed is in the construction of undersea oil rigs. Instead of developing installations that are easier to assemble and build in position, one school of thought has designed a rig that will move from place to place as the oil wells dry. One of the two movable rigs in the world has just completed trials in the Sea of Japan and is now in operation. The *Ocean Prospector* is a giant platform, 344 ft × 263 ft, sitting on sixteen hollow columns, which in turn are connected to four horizontally-placed buoyancy tanks. Four large cylindrical tubes connect the four buoyancy tanks in transverse fashion, thus acting both as the main structural interconnecting members, and as stabilising elements. The main deck, covering over 28,000 sq. ft, sits 120 ft above the tanks. During the cruise from one site to another the rig rides high on air-filled tanks. Once at the site the tanks are partially filled with water lowering them to a depth of 70 ft below the waterline, leaving the main deck 50 ft above the waves. The rig is propelled by two screws mounted on the aft end of two of the buoyancy tanks. These give *Ocean Prospector* a speed of 7 knots forward and 3 knots reverse. Steering is said to be excellent, and the rig can turn through 180° in two lengths. On site, chains on 15-ton anchors keep *Ocean Prospector* in position. The Japanese designers claim that even in the heaviest seas drilling operations will be possible, since the rig, because of its structure, presents little or no front to the surface waves, while benefiting from considerable inertia from the underwater crossbeams.

Offshore Mercury is the other ocean-going oil rig. The design is certainly more revolutionary than the Japanese structure: the *Mercury* deck sits on a huge shallow draught hull with an overall length of 276 ft and a beam of 130 ft. Propulsion is by twin diesel-electric screws and she can sail for 45 days, or 7000 miles, without need for replenishment. Her cruising speed is expected to be about 7 knots. The main deck includes a heliport, and 3500 tons of stores and equipment can be put aboard. But the unusual thing about *Offshore Mercury* is the fact that, once she reaches the drilling site, she lowers her own support legs to the sea bed. These legs are in fact four enormous derricks, one at each corner of the deck, slotted into their own winding gear. In the centre, between them, is the oil derrick.

Whichever of the two movable rig designs turns out to be the better, the whole approach indicates a considerable increase in the number of undersea wells being sunk in the next few years. As it also looks increasingly probable that oil deposits lie in most of the continental shelf areas of the world, the day may not be far off when the developed nations of the West, and in particular Europe, will be very much less dependent on supply from the Mid-East countries than they are today. It is interesting to speculate on the political ramifications of such a change. Indeed there are those who believe that the present hard line being adopted on oil prices by the Arab countries is no more than an attempt to cream what profits they can before they lose their pre-eminent position as the source of much of the world's fuel oil. In anticipation, designs are already being studied for offshore ports: the days of the small cargo-carrier are undoubtedly numbered as containerisation develops, and as the ships get bigger, both the navigational and environmental risks grow so long as the giants are forced to put in to coastal ports. The Zapata Norness company in America is now considering building a vast terminal in the estuary of the Mispillion River running into Delaware Bay. The terminal would be built to handle export coal, and eventually imported iron ore. According to a representative of the company there will be, by 1974, no less than 700 ships exceeding 250,000 tons in weight in operation round the world, none of which could use any port on the Eastern American seaboard when fully loaded. The need for a terminal, the designers say, is obvious: 'It is inconceivable that the US should not participate in this developing ocean transportation revolution, since the very competitive position of the United States is at stake.' The terminal would be located on a 300-acre island composed of material dredged from the Delaware Bay itself, and would sit about 3 miles offshore. The location was chosen because of its strategic position in relation to the American mid-Atlantic coast industrial area. Once the super-carriers arrived, their load would be automatically transferred to small, 40,000-ton shuttle barges, to be taken into a coastal port nearby. These barges would also bring out and automatically load material for export. The entire project is under review from an environmental point of view. What the construction of this size of artificial island will do to the ecology of the entire Delaware Bay area is not known. Conservationists have expressed fear that it would totally alter the balance of undersea life in the Bay, and that the effects would spread to the Atlantic in the immediate area.

There is no doubt that over the next decade technologists are going to be able to provide the means for us to begin extracting minerals from the ocean on a vast scale. Such an increase in natural resources flooding into the industrial complexes of the advanced nations may have repercussions that could change the political map of the world. Any country whose economy depends at present on the fact that it possesses a giant deposit of one particular mineral – like Zambian copper, Kuwaiti oil or South African gold – could find itself bankrupt shortly after the discovery of other equally large deposits of that mineral in the continental shelf of some other nation. Whole areas of the world, whose political allegiance is important to one of the Super Powers because of rich deposits within their boundaries, might have their entire political and social development arrested as they found themselves becoming unimportant economic backwaters overnight.

However, these changes are nothing compared with the effects undersea mining and activities associated with it might have on the ecological balance of the oceans themselves. Nobody knows what will happen if we alter the mineral content of the sea and change the shape of its floor. The way in which the ocean behaves may depend on those two factors. Deepening one area and filling in another would almost certainly redirect the movement of the water, bringing a change in temperature which would in turn affect the climate of the nearest land mass. Most important of all, the effect on life beneath the surface could be disastrous. The ocean contains not only the mineral resources we need if we are to continue to have the standards of living associated with being an advanced industrial society, but much of the food we need if we are to support a rising world population. The decisions we make in the next decade on how and to what extent we can permit exploitations of the world's oceans, whether on a free-for-all basis, or under the control of some kind of international rule of law, may be the single most important factor in determining what kind of future we have – politically, socially and economically – at the end of this century.

3   The Last Frontier

More than ten thousand years ago, if you wanted food, you went and looked for it. If you found it, in the form of edible plants or grazing herds of unsuspecting wild animals, you lived well until it either ran out or ran away faster than you could. When hunger returned you were obliged to go through the same process all over again. And when you failed to find the food you were looking for, you died of starvation. It was a very unsatisfactory state of affairs.

Then one day mankind decided to improve on the efficiency with which nature provided for him. He settled down, and began to grow plants, and to domesticate animals, so that whenever he wanted food, all he had to do was reach out and take it. The new way of living was eminently satisfactory, but it brought with it a new and pressing problem: it cut the death rate. With more mouths to feed, it became essential to extend gardens, corrals and pastures farther into the surrounding countryside, and make sure they were safe from passing tribes who were still eating food wherever they found it. As long as these strangers could be dissuaded from doing what they considered perfectly normal, and went on their way to look for food elsewhere, food stocks and the continued existence of the tiny settlement were assured. Thus it was that frontiers were born. As the settlements expanded into towns, and the towns into states, the frontiers were pushed out in order to provide more room for food production to support the growing populations.

Today there are no more frontiers to cross – on the land. But still the population grows, and with it the need for ever-increasing food supplies. So in the last 50 years we have been obliged to develop fertilisers and pesticides and improve our techniques to make the land produce more than nature ever intended. We have begun to experiment with artificial foods. But worst of all, we have crossed the one remaining frontier on Earth and pillaged with total disregard for the consequences. We have entered the territorial confines of the Earth's oceans. Unlike every other frontier we crossed in history, we have made no attempt to settle and develop it for the benefit of future generations. But it is beyond that last frontier that the future of the world's food supplies may lie.

Yearly it becomes more difficult to throw a net over the side of a fishing boat and haul in a worthwhile catch. For it is only recently that technology and a frightening acceleration in the growth of world population have combined to take more than the underwater environment itself can replenish. Every year the total number of fish in the world's oceans goes down. We are exhausting the sea's natural resources, just as we are those of the land. No amount of conservation can solve the problem without a parallel development in the production of protein-supply alternatives. Unless more effort is put into this area of research, it is forecast that natural fish resources will become critically low within the next 15 years. Meanwhile technology goes on developing better ways of catching what fish are left, on the grounds that we all still have to eat. And it is in honesty difficult to call this a short-sighted policy – we *do* all have to eat. The question is how long can we afford the luxury of natural seafoods?

There is a certain gastronomic irony in the fact that it is the French who have developed the latest method of making bigger and better catches. And they are using a very advanced version of one of the oldest fishing techniques known to Man: the mysterious and uncontrollable urge most fish seem to have to swim

Direction-finding systems that help divers working in total darkness. (See p. 73.)

towards a light source. The French Institute for Scientific Fishing call their new system 'Electrical fishing'. No net is involved. The technique consists of luring a fish towards a light and then into an electric trap. Using an 'electronic fish' the vessel searches the sea beneath it with an ultrasonic sweep. A fish shoal returns an echo to the operator on board. The ship casts anchor over the shoal and a lamp is lowered to sufficient depth to attract the fish. The trap, lowered just aft of the lamp, consists of two electrodes, one on the nozzle of a suction pump leading back up to the ship's hold, the other on a frame a further 20 ft aft. An electrical current is applied between the electrodes, charging a section of water about 200 sq ft in area. As the fish crowd towards the light the pulsating current stuns them and they are sucked up through the pump nozzle into the ship. The Institute says that most of their activity is directed towards the capture of the mackerel family on the grounds that these are the most plentiful. So far research has led the Institute to believe this system could be more efficient and less costly than conventional net-fishing methods.

One of the reasons the net has been of little scientific value to marine biologists is that there is no means of knowing precisely at what depth the fish in a haul were caught. Even the sampling net at present in use cannot be lowered to a point whose depth is known exactly enough to provide the detailed information the biologists need. Now, however, the British National Institute of Oceanography has solved the dilemma. Their new net carries a pressure-activated acoustic device to warn the researchers when it has reached the depth required for investigation. The device is called a 'pinger', because it emits two acoustic signals (radio waves will not travel well under water). One signal comes every two seconds; the second is triggered by the pressure sensors and is calibrated so that below a preset depth it will sound a given number of tenths of a second after the master signal for every given number of feet it descends. By measuring the time interval between the master pulse and the secondary signal, the biologists can work out with extreme accuracy where their net is. The point of being able to alter the pressure sensor to trigger the secondary pulse only within a given depth band is that the same equipment can be used at varying depths simply by calibrating it differently. Once the biologist decides his net is at the required depth he sends out an acoustic command which triggers a drop-bar. Fixed to the front of the net

Underwater spacesuit. Based on the principle adopted by the designers of space-suits of using the breathing gas supply to help insulate the body, this is the latest in diving suits. Air - from the diver's tanks not only keeps him alive and warm, but can be used to inflate the suit to a required level of buoyancy. This makes it easy to stay at a particular depth. Returning to the surface with a load is easy — inflate more, increase buoyancy, and the suit takes the load.

Bridge under water. This is a model of the British design entered for the Messina Straits Bridge Competition, sponsored by the Italian government. The Straits run between the toe of Italy and Sicily. The bridge is in fact three underwater tunnels, towed into position by tugs, sunk to the required depth, and moored on high-tension hawsers just below the depth at which they would be affected by surface wave movement. Two tunnels carry road traffic, one rail; each tunnel has a section under the road or rail surface, for services like light, gas or electricity.

The first production model of the US Navy's 'Universal Capsule', NEMO, being tested off the Bahamas. The sub is planned to provide data on sea-floor construction, underwater flora and fauna and diver behaviour. NEMO is capable of diving to over 1000 ft and needs only one support ship with a crane capable of lifting 15 tons.

This sampling net, developed by the National Institute of Oceanography, can be lowered to a predetermined level for sampling of specific water layers. The opening and closing of the net mouth is remote-controlled.

Bottom of page: This semi-transparent shrimp is a member of the commonest species in the oceans. Researchers want to know more about these creatures because they could be useful in large-scale fish farms. Big fish eat little fish, and tender shrimps like these might be used to fatten shoals in commercial fisheries.

in the closed position, the bar drops, and the net is open, ready for sampling. The range at which the system will work extends to about 3 miles, but there is no reason to suppose it could not be made greater with more sensitive hydrophone equipment.

Of all the efforts being made by marine biologists to unravel the secrets of the ecological chain of life under the sea, perhaps the most fundamental is that going on at the Scripps Institution of Oceanography in San Diego. There a team of investigators have succeeded in duplicating in the laboratory the entire life cycle of a tiny sea shrimp called a copepod. Their research has shown that this tiny animal is probably the very first in the chain of 'who eats who' that ends with the fish *we* eat. If the chain can be fully understood, then we stand a chance of being able to reproduce it artificially. If that can be done, then the control scientists would have over the availability of natural foods from the sea might add significantly to the world's larder.

It appears that the copepod is the only animal small enough to feed off the microscopic algae that abound in the ocean. The algae are converted into wax by the animal's metabolism. There are so many of these tiny shrimps it is estimated that the population off the coast of California contain in their bodies some 800,000 tons of liquid wax – far more than the *Torrey Canyon* and Santa Barbara oil spills combined. And this is very special wax – so special that it may be the principal source of energy in fish. Further along the life chain the copepods are eaten by predators – anchovies, sardines, herring, young salmon, etc. These in turn convert the wax back into fats so widely used for human foods. The copepod starts

Shark fishing by sub. This 16½-ft tiger shark was shot from a wet submersible during a shark tournament at Palm Beach, Florida. Vehicles like these can be used as underwater 'taxis' for frogmen. When submerged, they fill with water so the crew must wear breathing apparatus.

it all off in what for it is a big way, because the well-fed copepod contains as much as 70% of its own weight in wax. It is almost wholly devoted to producing the first, most fundamental food in the ocean.

The kind of work being carried out on the copepod will be of tremendous benefit to other researchers who are working on ways to increase the supply of fish through fish-farming. This is one of those areas of marine research that has been dogged by setbacks and unrealistic expectations. In the past few years it has been realised that most fish-farming attempts suffer because of the difficulties in getting the temperature of the water and its circulation exactly right for every stage of the fish's growth. Cannibalism has on many occasions wiped out up to 30% of a tank population. Apparently rearing the fish and letting them loose in the sea has not produced anywhere near the increase in stock that was expected. All in all, the only immediately viable proposition lies in the farming of the luxury fish foods: salmon, white fish, shrimps, eels, crayfish and oysters.

The Perry Shark Hunter is primarily a recreational vehicle although it can be used for some scientific purposes. It is designed to operate at a depth of about 100 ft. The controls are similar to an aircraft; a joystick activates two external diving planes and foot pedals control the rudder. Power comes from a 4½-hp motor and it has two forward and reverse speeds.

The most successful breeding so far has been with shrimp and prawn. Japanese techniques being used in specially constructed shallow salt-water 'lagoons' in Florida aim at producing enough shrimp to put 20 million into the sea and have enough left to make a profit. The system is extraordinarily simple. Under natural conditions less than 1% of all baby shrimp survive. By rearing them in protected conditions over fifty times as many survive. As for maintaining the water at the correct temperature, one shrimp farmer in Britain has solved the problem by setting up his farm in the hot water effluent from a nuclear power station. Here the shrimps grow to full, marketable size in half the time it would take under natural conditions. All it takes to run a successful operation is care. The mother shrimps are kept in plastic tanks, where they lay their eggs. The eggs hatch into larvae, three or four thousand per mother. These larvae are then carefully filtered through a very fine mesh and collected in a plastic bowl. They are then fed on an even smaller, brine shrimp, which is brought in dehydrated form from California. As the baby shrimps grow they begin to exhibit one trait of character that can drastically reduce the number of shrimps the farmer will eventually sell: they eat each other. In the shrimp farm this cannibalism can be reduced by inserting horizontal barriers in the shrimp pens, which by segregating the shrimps into small batches cuts down on the crowded situations that give rise to cannibalism. The success of the shrimp farms so far has been so spectacular that some experts predict battery-reared shrimps and prawns could replace all other fish in the human diet over the next 20 years as the fish available in the ocean decline in numbers and the shrimp farms multiply.

One other switch in taste we may all have to accept is the move to oyster-eating. One day they may be the only readily available alternative to shrimps and prawns. The reason oysters may soon no longer be the delicacy they are today is, oddly enough, due to our increasing pollution of the ocean. As industrial waste, offshore drilling and the diseases they helped cause gradually overwhelm the natural oyster reserves of America, and in particular the Delaware Bay beds, researchers have begun an intensive survey to try and save the industry. Their work has shown that the intensive breeding of oysters on a factory production-line basis is not only feasible but could turn out to be profitable. The entire research project that came up with this happy result was carried out on a computer, programmed with every single fact relevant to the rearing of oysters. Every stage of the breeding process was then processed in digital form and entered into the computer's memory banks alongside the oyster's facts of life. The computer analysed the information and suggested a production process. It's production-line oyster factory consists of three sections: a central food supply would be grown in the Algae Section. Algae is the oyster's staple diet at all stages in its life. As through the rest

Shrimp farms. The mother shrimps are kept in plastic tanks where they lay their eggs.

of the factory, the computer suggested keeping algae production going on non-stop, 24 hours a day. The second main area would be the Hatchery, the life centre of the whole complex. Here the oysters would be spawned and set, before they were placed in growing tanks for the rest of their 12-month life. Oysters readying for spawning would be kept in cool water at about 20°C. Over 7 to 10 days the water would be slowly warmed to 25°C., at which point, according to the computer, water 2° or 3° warmer should be added, and spawning would result. The larvae would then be held for 24 hours before transfer to another part of the Hatchery, the development tank. Fifteen days later they would be moved to the setting tank to grow until the required number of oysters per square inch was reached. Three months of growing in this mathematically controlled environment, and the computer said they would be ready for their last move, to the factory's third sector – the Growing Tanks. Here they would be stacked either vertically in deep water for high-level production, or horizontally in shallow water. The entire oyster factory would work 24 hours a day throughout the year – and with that would come the end of expensive and seasonal oyster-eating. If the computer is right, only one problem remains – persuading people they actually like oysters.

In spite of setbacks, the Americans are more confident than most about the future. Recently the Director of the Ocean Foundation of Hawaii predicted fish farms operating 1200 ft under the Pacific, with algae-grazing fish being herded by dolphins. Unfortunately dolphin experts tend more and more to think that the mammal in question is a good deal less intelligent than was previously thought. Be that as it may, the prediction included an opinion that Puget Sound, with an area of 60 × 15 miles, could alone yield fish equivalent to the present entire United States catch. The forecast involved giant sea-floor fish habitats, like corrals, tended by farmers working in 3-man submarine capsules. The likelihood of this dream ever becoming a reality depends largely on whether or not we do two things: stop taking fish out of the ocean faster than it can restock itself; and cease polluting the water at the rate we are doing so today. Pesticides washed out to sea kill the algae, the tiny plants in the ocean that are the staple diet of many fish, and in doing so drastically reduce the amount of oxygen (which the algae produce)

In the still, calm waters of Ardtoe in Scotland – a tiny experimental fish farm. Within a few years rearing fish in captivity could become a multi-billion-pound world-wide industry.

present in the sea. That oxygen is also vital to fish. Both overfishing and pollution have helped to reduce the number of fish in the seas by a tremendous amount. Already scientists are looking for ways of increasing the number so that we have enough both to eat and to stock the ocean.

The most spectacular attempt so far has been to use radioactivity to double exactly the number of fish that would normally be produced by one batch of fish eggs. The fish involved are plaice, a hardy stock, that take to forced breeding fairly easily. They are already a popular food, so there is no need, as often occurs, for the scientists to persuade the consumer that just because it's artificially plenti- ful, the new protein-rich source of food is palatable! The researcher working on the plaice takes his specimens from the trawlers coming into Lowestoft, in Britain. Back in the laboratory the sperm, or milt, of the male plaice is removed, and chilled on ice to await the arrival of female eggs. Meanwhile it is sealed in an irradiation chamber and saturated with gamma rays from a radioactive Cobalt 60 isotope. The gamma rays smash the male chromosomes in the milt, without, however, depriving it of its ability to fertilise the female's eggs. The reason for this destruction is that by eliminating the male chromosomes, the breeder is doing away with any extra hereditary faults the chromosomes might carry. It is, in a sense, quality control, because the breeder knows that the baby plaice will have exactly the same characteristics as their mother, uncomplicated by those of their father. The female plaice can produce anything up to fifty thousand eggs, which are gently squeezed from her egg sacs into a dish of sea water. The milt is taken off its ice and put into the same dish, where it heads straight for the female eggs, fertilising about 90% of them. Because the male chromosomes have been destroyed, the embryo about to form contains only two female chromosomes, one of which would naturally separate and die within 20 minutes. But because the fish needs two chromosomes to survive normally, the female chromosomes must be induced to join together. This is effected by syphoning the eggs into freezing cold water 15 minutes after fertilisation, as the sudden lowering of the temperature causes the two female chromosomes to shock-join. After 5 weeks of steady growth and careful feeding, the small fish reach metamorphosis, when they change into the familiar oval flat-fish shape of the plaice. After a few generations of these females have been fertilised by male eggs whose chromosomes have been destroyed by radiation, all the baby fish look exactly the same. They are the result of the latest in selective breeding techniques. Each one is healthy and up to good food specifications. Hopefully the breeder will be able to produce so many of them under these carefully controlled conditions that half his stock will be dumped in the sea to replenish the dwindling resources that sonar equipment reveals almost everywhere throughout the oceans.

In spite of all the technology that reveals the wonders of the sea to us – via umbilical cord, or mechanical grab, or echo-sounder – and in spite of the in- creasing understanding science gives us of the workings of the oceans themselves and the life within them, the one fact that ultimately attracts man to the sea is that he does not belong there. For centuries his growing knowledge has given him mastery over the mountains, the jungles, the most inaccessible land areas of the earth, and even the sky and the fringes of outer space. Yet in all those centuries of progress he has managed to penetrate no more than a few hundred feet below the surface of the ocean, and then only for fleeting moments. Even the most sophisticated and powerful tools we have at our disposal have not been able to show us even what it looks like at the bottom of the deepest sea. The waters of the Earth are so much more alien even than outer space. There all we need to survive is a thin shell capable of withstanding the meagre pressure of the oxygen our weak, vulnerable bodies need to breathe. At the bottom of the ocean we need a shell stronger than Man has yet made, and on the way there and back we must take infinite care, go terribly slowly, if our bodies are to survive the immense pressure changes. It is perhaps because Man has lived for so long so close to this totally hostile world, a world that if he were unprotected would kill him within minutes, that with his new-found technology he wants to master it. For it is the last unexplored frontier on this planet, and the more we test ourselves against it, the more challenging the enterprise becomes. But in 40 years of experimental diving, the deepest a man has managed to go and survive is 820 ft.

Before we inch a little deeper into the sea, we must first find out what will happen to us by going 'deeper' on dry land. These 'dry dives', as they are called, take place in laboratories under the most stringent safety precautions. The divers 'descend' to the simulated depth in hyperbaric chambers, small spherical or cylindrical structures of steel capable of withstanding the enormous pressures that are built up inside them to reproduce conditions hundreds of feet below the surface of the sea. No dry dive ever takes place without a team of highly qualified engineers and doctors present. We still know very little about what happens to the human body as it goes through the process of diving deep, and the procedure is painfully slow. Any sudden change in pressure could be fatal to the human guinea-pig inside the chamber. It is for this reason that the latest deep dry dive, carried out by the French in a test programme code-named Physalie 5, took 5 days to complete. And the two divers only reached the equivalent of a mere 1706 ft deep. But that was a world record.

Patrice Chemin and Bernard Reuillier, the record holders for the world's deepest 'dry' dive. They reached a depth equivalent to 1,706 ft.

The dive took place at the COMEX Hyperbaric Research Centre in Marseilles. The divers were wired up to a battery of monitoring devices. Encephalograms measured brain activity; records were made of second-by-second changes in cardiac, muscular and respiratory behaviour; tremometers were taped to their hands to record the slightest trembling in the fingers; miniature transducers helped to pick up minute alterations in the pressure and flow of their breathing; psychomotor tests were carried out periodically to check their reflexes, manual dexterity and intellectual agility. Diving began on a Monday, with a very slow build-up of pressure in the chamber to simulate a very slow rate of descent. With frequent pauses for acclimatisation, it took 2 days to get the divers 'down' to 1510 ft. They had been to this depth before, but the next part of the descent was critical. On previous dives the researchers had observed the development of a mysterious and dangerous reaction in the nervous system of a diver under very high pressures. It appeared that, under such pressure, the body tissues became saturated with the gas being breathed in the chamber, and this in some way caused changes in the way the nerves work. The principal symptom was severe disorientation – the diver hallucinated, forgot where he was. There was a danger that he might even take off his breathing mask. That is why these dry dives are essential forerunners to attempts to go deeper in the ocean. In the hyperbaric chamber, if something goes wrong, the diver in trouble can receive specialist treatment very quickly.

After a night's rest at 1510 ft, the dive continued until, in the afternoon of the following day, they reached the record depth of 1706 ft. At that point, equivalent to the very uppermost levels of the ocean (there are parts of the Atlantic that go down to over 30,000 ft) the divers were experiencing the effects of unbearable pressure, at over 110,000 lb per square ft. Almost as soon as the deepest point of the dive was reached, the ascent to the 'surface' began. It, too, took 2 days, as their bodies went infinitely slowly back through the pressure changes to normality.

The dry dive provides important data about the changes in the physical condition of the divers involved, but does not actually subject them to immersion in water. It is, in a sense, a run to test theory under controlled conditions. The next step is land-based 'wet' experimental dives. The 'wet' dive record was recently set up in America. Once again it took place in a chamber, made of steel.

The record 'dive' was made inside these hyperbaric chambers at Marseille. The two men lived in them for four days while the pressure was steadily increased to 110,000 lb per square foot — and then slowly brought back to normal again.

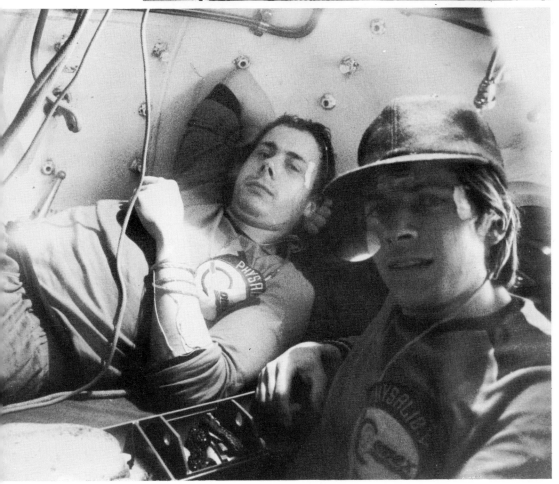

This time, however, the chamber was divided into two sections, one 'dry', where the simulated descent was made, and the other 'wet'. This second section was connected to the first with an airlock, and was filled with 7000 gallons of water saturated with 8500 lb of sodium chloride to simulate sea water. Five commercial divers from the Taylor Diving and Salvage Company spent 20 days in their steel chamber going through a descent to 1000 ft that was divided into five stages, each one of 200 ft. At each stage the build-up of pressure in the dry section was halted, and one by one the divers went through to the 'wet room'. There, wearing suits and masks, they carried out a series of tests to determine their ability to work at that particular pressure level. Each time a diver entered the wet room, electrodes pasted on his skin, connected to wires leading through an umbilical cord to the experimenters outside, transmitted information about all aspects of his physical state. As the dive went deeper the ability to stand work loads was closely monitored with a unique device, called a trapeze ergometer, designed to impose a strain on the diver equal to that of swimming against a 1 mph current.

These two dives were so successful that the experimenters, both French and American, are now talking euphorically about getting a diver down to 2000 ft,

294 ft deeper than the present dry dive record. It is a measure of how immensely difficult the task is that another 294 ft should generate such world-wide excitement.

The men involved in these experiments are a long way ahead of the rest of the world's divers, most of whom work at depths that rarely exceed 300 ft. And at this relatively problem-free depth, technology continues to make the diver's life easier for him. One of the ever-present hazards of working underwater is the 'weather' down there. A change in current can stir up mud and silt from the bottom, reducing visibility to a few feet at the same time as it sweeps the diver away from the area he may want to work in. Without a view of the bottom it is all too easy for a diver to lose his sense of direction. That is one of the reasons why even the most experienced inhabitants of the new underwater habitats, like the one in the Caribbean, will venture only a short distance away from their submarine homes. And these conditions often make the business of getting down to the sea bed to investigate a site or a wreck too difficult to try except in the clearest of water.

A French company, the Thomson organisation, have now developed a direction-finding system that will permit divers to work in total darkness. Once the wreck or site has been located by the survey ship, a small epoxy resin cylinder is dropped overboard on an anchor. The cylinder contains a tiny transistorised acoustic 'pinger'. The device sends out a strong signal in all directions through the water, with the strongest propagation horizontally. Aboard the survey ship they know from the length of line paid out to the pinger what depth it is sitting at, so all the diver has to do is descend to that depth and then start to search for the pinger at the level where he is likeliest to receive the strongest signal. His personal receiver is a slim baton, held in the hand. As he swims along with the receiver-locator held out in front of him, an acoustic receiver buried in the lower part of the baton listens for the pinger's signal. When it picks up the signal, it can measure the strength of reception in order to identify whether it is coming from the right or the left of the diver. Then it activates one of two light signals on the top of the baton. All the diver has to do is follow the light until he locates the pinger, then follow the pinger's anchor line down to the sea bed.

Thomson has also developed a portable sonar system which a diver can use to find a wreck or any object on the ocean floor, once the survey ship's own sonar has located its general position. With this system a diver can drop over the side and go straight to the site no matter what the visibility underwater. The apparatus looks like a small searchlight with a handle on either side. It is transistorised and completely self-contained, so there are no lines back to the surface to bother the man using it. The sonar sends out an ultra-sonic acoustic signal up to 300 ft ahead. When the signal strikes an object it bounces back, just as an echo does. The returning signal mixes with, and modulates, the outgoing signal, changing its pitch. The diver hears it through a pair of earphones, and the level of pitch tells him how far away the object is. The sonar can also be used in conjunction with the pinger. In this case the pinger's signal mixes with the sonar's to tell the diver in what direction and at what distance the pinger is located.

These two new developments operate in the acoustic range, because so far it appears that only ultra-low-frequency radio waves will penetrate water, and at that frequency extremely long aerials are needed for reception. Recently revealed military experiments suggest that this method is being used for communication between submarines on patrol underwater. Obviously this technique involves equipment that no individual diver could handle, so scientists have concentrated on the acoustic frequencies in their search to find a way of communicating with a free-swimming diver.

Another new development looks like ending the isolation in which most divers have had to work. Divers will no longer have to surface to report on their work and

Michael Liogier and Christian Cornhllaux on the way down to the deepest sea dive yet. They reached 820 ft.

Working 820 ft below the surface. No one had ever done it before. (See page 70.)

Portable sonar equipment for locating wrecks on the ocean floor. (See page 73.)

receive decisions about what they should do next – and that saves money. The new device is an electronic converter which turns the spoken word into acoustic signals. It is positioned, together with a tiny microphone, in the diver's breathing mask. The signals are boosted by an amplifier and transmitted through the water to a receiver hung over the side of the surface vessel, or indeed fixed on the outside of an undersea habitat or submarine, at any distance up to 18 miles away. The receiver contains an 'unscrambling' unit which converts the acoustic signal back into speech.

Until this new device goes into general use, divers will continue to communicate with the surface through conventional telephone lines. Apart from limiting the diver's freedom of movement, these lines provide the compressed air that divers usually breathe, but below 600 ft another problem arises. At that depth the intake of high-pressure air containing principally oxygen and nitrogen occurs more rapidly than the body can take. The compression forces the gases through the lungs into the bloodstream at a potentially poisonous level. In answer to this problem, scientists discovered some time ago that if a deep-sea diver breathes a mixture of oxygen and an inert gas the toxicity of the high-pressure oxygen is substantially reduced. The inert gas predominantly in use is helium, and its effect is to 'water down' the rich oxygen. One symptom of inhaling helium is to distort the voice of the diver into an almost unintelligible duck-like sqawk, because the gas compresses the frequencies of speech as much as 2·7 times. The extent of this compression is related to the amount of helium in the mixture, which in turn is related to the depth of dive. The result is that some elements of speech, the fricatives and sibilants (consonants like f, t, k, s) and the voiced sounds (the vowels and some of the other consonants produced by resonance in the head: m, b and n, for example) sound higher than they would normally, because sound travels faster through helium than through air. What the new device does is to hold these sounds in 'store' for 2·5 milliseconds, and then stretch them. The

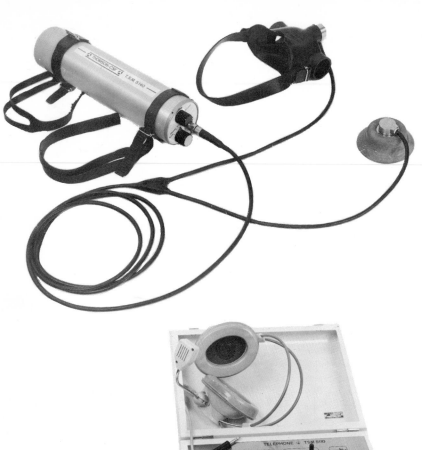

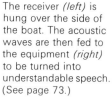

Microphone and acoustic transmitter designed for frogmen. Equipment like this does away for the need for underwater telephone cables.

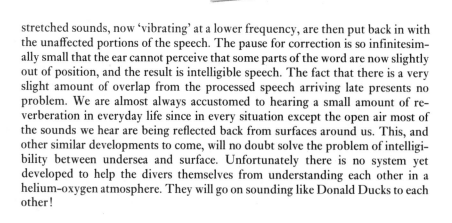

The receiver *(left)* is hung over the side of the boat. The acoustic waves are then fed to the equipment *(right)* to be turned into understandable speech. (See page 73.)

stretched sounds, now 'vibrating' at a lower frequency, are then put back in with the unaffected portions of the speech. The pause for correction is so infinitesimally small that the ear cannot perceive that some parts of the word are now slightly out of position, and the result is intelligible speech. The fact that there is a very slight amount of overlap from the processed speech arriving late presents no problem. We are almost always accustomed to hearing a small amount of reverberation in everyday life since in every situation except the open air most of the sounds we hear are being reflected back from surfaces around us. This, and other similar developments to come, will no doubt solve the problem of intelligibility between undersea and surface. Unfortunately there is no system yet developed to help the divers themselves from understanding each other in a helium-oxygen atmosphere. They will go on sounding like Donald Ducks to each other!

Ever since the invention of the wet suit for divers, various methods have been tried to increase the insulation of the wearer against the heat loss he suffers while swimming in water cooler than his body temperature. Up to now the most successful idea has been to allow a small amount of sea water to enter the suit, to be warmed by the body and act as an extra protective layer. Now a Swedish design may have made that unnecessary. It's called the Unisuit and in some ways is similar to the suit worn by astronauts. Like the spacesuit, a zipper, in this case running all the way down the back between the legs and chest, divides the suit in two for ease of donning. As with the Apollo garment, the Unisuit is slightly inflated by the diver's own air supply which provides an insulating layer of warm air. The result is that the wearer can work for much longer periods in cold water, as trials in Arctic seas have proved. Again, like the spacesuit, this one is meant to be worn with varying thicknesses of underwear according to the temperature outside. But it is the flow of air in and around the suit that is the most valuable new facility. Pressure inside the Unisuit can be increased or decreased to maintain negative buoyancy, and save the diver the effort of having to swim to remain at any given depth. Increase the pressure still more, and the suit becomes a work aid in itself. When you have to move a heavy object underwater, or bring it to the surface, you simply take hold of it, allow the internal suit pressure to build up, and the extra buoyancy lifts you and the object off the sea bed.

But despite developments like these, one thing is becoming clear. It begins to look as if the oceans are too hostile ever to permit men more than brief, dangerous visits. All the highly coloured predictions of Man living and working under the sea as he does on the land and in the air fail to take account of the simple fact that this is one element to which Man cannot adapt himself. His very physical structure is wrong. Other air-breathing mammals, like monkeys, have been submerged to depths of 3000 ft, breathing mixtures of inert gases and oxygen, and the disappointing results of those experiments only confirm how tremendous the physiological and psychological stresses are. They show that the biomedical problems affecting the survival of a diver are daunting. The most obvious and immediate are those that come from working in a wet, cold, dark, crushing liquid. Breathing air under pressure can turn both the oxygen and the nitrogen in it into potentially lethal gases. The nitrogen can cause nitrogen narcosis in the brain, and bring on loss of consciousness. High-pressure oxygen can be equally poisonous. At greater depths it is difficult to breathe even at rest. In addition the loss of body heat is considerable due to the cold water surrounding the diver, and the fact that the compressed gases he is breathing are also cold. The full effects of high pressure applied all over the body are not yet known, but early tests show them to be dangerous; and should all or any of these conditions develop into an emergency situation, the diver will die if he attempts to reach the surface too quickly.

There are also other problems to be found. The psychological effects of living in very crowded conditions in an undersea habitat indicate that the inhabitants may have to be as carefully chosen as are the astronauts now. No one yet knows if the body needs special kinds of food and drink at great depth. Should one of the divers fall sick, tests have yet to be made to determine how the body will react to drugs. For some reason the brain's activity at great depth is inhibited. Pressure forces the body's joints to grind and rub together, and the ball joints at hip and shoulder can become flattened and induce bone necrosis. And far from adjusting to conditions below the surface, the chest muscles become more and more incapable of forcing the chest wall in and out during the breathing process.

It is this last factor that imposes the greatest limitation on Man under the sea: his lung function is inadequate. Dr L. West works for the American National Aeronautics and Space Administration on the problems of lung behaviour in

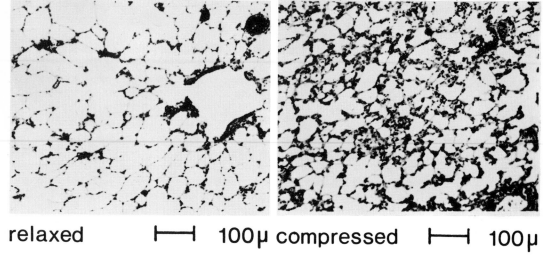

relaxed ⊢—⊣ 100µ compressed ⊢—⊣ 100µ

the zero gravity state encountered during long flights in space. As part of the research, he and his team have carried out studies of the diving mammals, notably the sea-lion. They have found that the reason such animals do not get brain narcosis from having excess nitrogen forced into their bloodstreams under pressure, or the 'bends' as the nitrogen in the blood then expands into bubbles as the animal returns to the surface, is because their lungs are unlike ours. Their main passages within the lung are heavily protected by muscle – almost 'armoured' – right to the point where the air passes directly into the blood. In all mammals, including man, the air permeates the walls of tiny sacs (called alveoli) deep in the lungs to pass directly into the bloodstream. When a man is at depth, the sacs in his lungs retain their normal shape, permitting the high-pressure gas to enter the lung and leave through the sac walls into the bloodstream. In sea mammals, the tiny alveoli collapse, sealing off the bloodstream from contamination by the air held in the lungs under pressure. This research confirms the fact that, unless some way is found to restructure the human lung, man stands little chance of spending much time beneath the ocean surface, except in heavily armoured tanks.

Apart from his lung shape and function, man is also the wrong *shape* to work efficiently in water. Another American scientist we spoke to, Dr D. Denison of the University of California School of Medicine, took up the problems of man's subaquatic inadequacies when he speculated on ways in which man might be re-designed to overcome some of those drawbacks. His idea is to attempt to alter the shape of a man to resemble more closely one of the fastest mammals in the sea: the dolphin. Here is part of the transcript of our interview with him: 'The reason why a dolphin can swim so much faster than a man is largely because of his shape. The problem with an irregular body when it is pulled through the water is that the water doesn't follow the boundary of the object all the way round, and the end result is that behind the body there is dragged a long tail of dead water.

'Now, there are three ways round this problem. You can reshape the body so that it has a tapering end, and the water will then follow it around. A perfect body of this kind is called a laminar spindle. The finest examples known in nature are the dolphin and the tuna fish, and of course that is the reason why many ships and planes are built along similar lines. Another thing you can do is to put spoilers on the back that disrupt the flow and so break up the column of dead water. The third method is to put a fan at the back and actively propel the dead water away. Of course fishes and dolphins do all of these. They are all shaped more or less like

Lung of dog.

The microscope reveals why men – and dogs – cannot normally survive the rigours of deep diving.

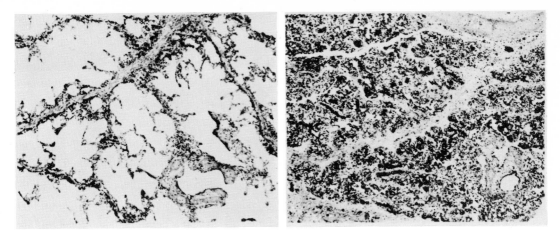

relaxed   ⊢ 100 μ   compressed   ⊢ 100 μ

Lung of sea-lion.

laminar spindles; they've all got spoilers; and they've got fans at the back. A dolphin is exceptionally well-designed in the essential modifications in that mouth and eyes and stabilising fins are all in one horizontal line, so that they generate only a single cone of that turbulence. This greatly improves their performance underwater.

'However, if you were to get a model dolphin, a highly polished aluminium dolphin, the best that a man could prepare, and tow it through the water, you would find that it would still require five times as much power as was available in the dolphin's body. Relatively recently it was established that the reason that the dolphin can move so much more easily is partly because of the construction of its skin. It contains many tiny canals, and any pressure wave that builds up on the surface is dissipated or damped out by the movement of the fluid from one canal to another. If you get an aluminium spindle and coat it with a dolphin's skin, you will find that the resistance to motion drops something like five-fold.

'Now, at first sight it seems ridiculous. How can we improve a man's performance underwater, looking at the dolphin as an example of a mammal? Well, the first thing that you can do is to reshape him like a laminar spindle. Unfortunately, man is built wrongly. His widest diameter is his shoulders, and they are rather forward. We can do nothing about pulling his shoulders further back along his body, but we can extend the front of his body and reshape the head. We could, for example, put his head in a perspex cone, and extend the cone way in front of his head, thereby effectively putting his shoulders halfway back along his body. And we could use that perspex cone to project information to the diver on depth, in a head-up display, compass-heading, the time that he has been down, and all these other important things that are required. At the same time we could refashion the rear part of the body by a properly-shaped wet suit, preferably of a less compressible material. His shape is now much more like a laminar spindle.

'The next thing we do is give him a tail, much more like a dolphin's tail. The dolphin proves to us that at least a tail of its design is capable of being moved through the water one or 2 ft at a time at one or 2 cycles per second, without generating turbulent flow and wasting effort. And if we know that when a man does this the great majority of his effort is dissipated in creating turbulent flow in the water, well, give him a dolphin's tail. Now, what else can we do for him? We can recruit more of his muscles to move that tail. This is already done to a small extent in the dolphin kick in ordinary swimming. But he needs to use more of the long back muscles. This is a simple engineering problem of giving him a transmission link between his body and the tail that uses the greatest amount of muscles possible. We could also pay attention to the other things that we have seen in the dolphin design, keeping essential modifications in a single line and perhaps giving him a coating similar to a dolphin's skin.

'Now, these ideas may seem far-fetched. You may think it is much easier just to give a man extra power in a pack, as a source of moving around. Nevertheless, the gains to be achieved from my sort of arrangement and improvement in efficiency are perhaps six- or seven-fold.'

Meanwhile the conventional test dives continue, pushing the frontiers back a few feet every year. But they go on slowly, in few centres and with few resources to back them. The dive researchers need much more money if they are to progress any faster, and the money is, as always, more readily available for projects that promise immediate profit. That is why the real future of undersea exploration lies with unmanned machines. They demand no complex built-in safety devices, and in the long run they cost less and do more. Indeed, the most sophisticated equipment being made available by the technologists, and the machines under study in well-financed projects, are almost without exception automatic.

Experts in diving techniques confidently predict that within the next 6 years men *may* be working on the continental shelves, living in habitats, drilling oil wells, operating terminals for oil and mineral storage. Yet the shelves form only a fraction of the undersea area to be exploited. And even at that relatively shallow depth the work will be a hazardous operation, liable at any time to run into danger. And in such emergencies the retreat has to be very fast indeed if the workmen are to survive at all.

In the light of the latest research, the plans being laid for underwater activity 10 years ago now sound very far-fetched. Then people were suggesting that undersea cities would solve the world's growing population problem by providing extra living space; the vast farms around the cities on the sea bed would feed their inhabitants; there was even an attempt to develop a brine solution saturated with dissolved oxygen which would fill a man's lungs, providing him with the oxygen he needed while putting an end to the problem of gas-filled lungs that could collapse at extreme depth. In spite of the technological developments over the past decade the difficulties still seem insurmountable. The sea has presented more, not less problems, the further the scientists investigate. Today it would be a brave man who would forecast that we will be very much more active, or efficient, or at home in the ocean's depths in 20 years' time. Quite apart from man's own physical and psychological limitations, even the considerable number of machines, manned and unmanned, that we have sent below the surface have told us very little. We still know virtually nothing about the fundamental question of how the ocean itself *functions*. Ocean dynamics is an infant science. But it has recently benefited from an increase in interest thanks to space exploration. As our techniques for examining the surface of the planet from orbit have improved, we have been able to analyse more closely the part the Earth's seas play in the life cycle of the entire planet. It is now recognised that the world's weather systems are almost exclusively dependent upon the interaction of the ocean with the atmosphere. As we look ahead to the possibility of a minute-by-minute world weather watch and perhaps, farther in the future, the possibility of weather control, the more we become aware of the need to understand the forces in those areas of the planet's surface covered by water.

A first step towards that kind of control is the construction of a 'model' of the oceans. One idea is that it should consist of a vast memory store in a computer – if one computer is big enough to hold it – containing every tiny detail of every single action of every part of the ocean at every foot of depth from the surface

A Tektite undersea habitat *(near right)* will sit on the floor of the ocean. It is one of the few underwater houses designed for work rather than setting specific undersea records. A design similar to this was used for two years in the Virgin Islands.

Flower Gardens Campus *(far right)* will consist of platforms raised above the sea. Laboratories will be standard modules that can be transported from the mainland by boat or helicopter. Living conditions for scientists who come here to work will be luxurious. There is one snag. The Gulf of Mexico is noted for its hurricanes. Should the weather get too bad all personnel will be evacuated to the shore. If there is sufficient time, the laboratory modules will be taken, too.

The Gulf of Mexico is a paradise for oceanographers. So American scientists are establishing the world's first open-ocean campus right in the middle of it — 120 miles from the mainland. It is called the Flower Gardens Ocean Research Centre and it gets its name from the coral reefs near which it is to be sited. The idea will be to provide faculties for scientific study of the ocean and the American Continental Shelf.

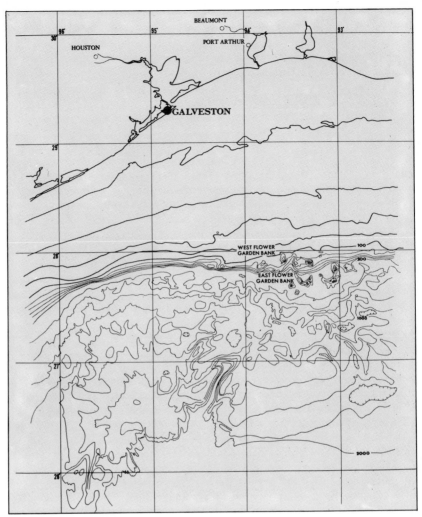

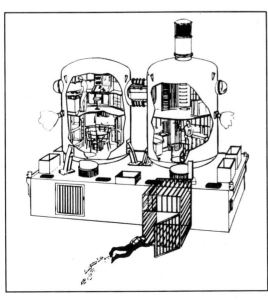

to the bottom. And there must be data of this kind referring to every single hour of the ocean behaviour throughout the year. When we have this computer model, we can begin to feed into it the plans we have for any kind of alteration of the natural state of the ocean: every ounce of material either to be taken from it or dumped into it; and every contruction or industrial operation planned to take place within it. After we feed the data in, the computer will show us in detail the effects, however far-reaching, of our actions.

Such a model is the dream of every oceanographer alive, but it may be a century before the day comes when we can construct it. But the first attempts to bring that day closer are already being made. At Harvard University two researchers have built a laboratory model of an ocean, in a very small-scale experiment to discover something of how the seas respond to the tremendous forces that set them in motion. The model is nothing like the computer version of the next century, but it is a good deal more sophisticated than the water-tank versions that have been attempted in the past. It consists of two circular blocks of clear plastic, one above the other, with a one-centimeter space between them. The upper surface of the lower block is spherically convex, and the lower surface of the upper block correspondingly concave. The one-centimeter layer between them is filled with water, and represents a section of the surface of a sphere, mounted in a near-vertical position to represent the orientation of the North Atlantic as seen from a point in space above the equator. This thin curving layer of liquid held in the plastic blocks is then mounted on a turntable to simulate the rotation of the Earth. As the model turns, the lower block can be spun by means of a motor and belt, to impart direction to the water layer.

Since the two plastic blocks are curved, it will be seen that the upper block represents the atmosphere above the water layer, and the lower block the Earth's crust beneath. So as the upper block spins it is acting on the water layer as the winds do on the surface of the sea. The wind is one of the two principal factors affecting the movement of the seas, apart from the effect of the Moon on the tides. The winds themselves come into existence as a result of the heating of the Earth's atmosphere. As the heated air rises, cooler air moves in to replace it, and in doing so displaces the surface layers of the ocean. At the same time the sun's heat also affects the water itself, by heating the surface, and causing volumes of water lower down to move in response to temperature gradients. These are very basic facts indeed, and yet to begin to trace the effects of the sun's heat much beyond these general trends is an almost impossibly complex task. The Harvard researchers are trying to test some of the general theories on ocean dynamics developed over the last 50 years. These theories say things like: 'If wind of a certain velocity is applied, a certain current will develop in a certain direction, given the rotation speed of the Earth.' The theories relate only distantly to the myriad dynamic movements that the ocean undergoes every moment of the day and night, but they may or may not establish very simple ground rules from which to begin experiments. The Harvard model may discover whether those ground rules are right or wrong.

Yet while these and other research organisations struggle on tiny budgets to carry out their investigations into how and why the ocean behaves as it does, how many fish there are left in the sea, how the ocean affects the very air we breathe – the rest of us continue to treat the sea as a vast garbage dump. Every year *millions* of tons of man-made wastes are poured into the sea without regard for the chilling fact that the sea is a moving dynamic body teeming with life that must react to the poisoning effect of this waste. We have no clear idea how the ocean will react in the long run, and yet we continue, even though the short-term effects are beginning to show themselves in frightening ways.

The first results of pollution have destroyed the myth that anything dumped in the sea undergoes dilution to an enormous extent, and that the dilution greatly reduces its harmful effects. The complex and efficient industrial processes that are present in the ocean, acting either through the living organisms present in the water, or the movement of the water itself, are brought to bear on whatever pollutants are introduced. Thus it is, for example, that birds found dying in convulsions off the coast of Holland had been poisoned by chlorinated hydro-carbons dumped many miles away over a year before. Tiny shrimp had eaten plankton affected by the chemicals, part of a pesticide discarded into the Rhine. Herrings and sand eels had later eaten the tiny shrimp, and had in turn themselves been eaten by the birds. By the time the chemicals reached the digestive systems of the birds, they had not been diluted, but *concentrated* many times. Each animal in the food chain that had eaten the chemicals had refined them many times over.

But it is not even necessary for a killer substance to pass all the way through the food chain in order to kill those waiting for the last meal in the queue. If one species half-way up the chain is wiped out, the rest of the members of the chain have to find alternative sources of food or they in turn will die of hunger. The human race is at the very end of every food chain in the ocean. The more we study our planet, and in particular the oceans that cover two-thirds of its surface, the more it becomes clear that Man's existence on Earth depends on the most immensely exact balancing of a number of inconceivably powerful factors: the exact and unvarying rotational speed of the Earth; its precise and unvarying distance from the sun; the exact and unvarying tilt of the planet on its axis. If any one of these change in the slightest degree, Man could be wiped off the face of the Earth. Almost certainly the killer on such an occasion would be the ocean. It could either freeze, evaporate, or move suddenly and violently. It is on the continuing stability of the ocean's behaviour that our lives depend, because as long as the sun continues to warm the Earth, a vital interaction between the ocean and the atmosphere will continue. Put very simply that interaction follows this pattern: the sun heats the sea, causing it to give off water vapour. The vapour moves with the winds and encounters land masses, which cause it to precipitate and fall to the ground as rain. The rain irrigates the land, permitting crops to grow. Change the chemical balance of the sea, and you change the rain and its effect on the crops. Once again the human race is at the receiving end of a chain of events which it is capable of disrupting through the effects of its own pollution.

But perhaps the most immediate way in which pollution can help us to commit suicide lies in its ability to affect phytoplankton – a plant so small it can be seen only through a microscope. The phytoplankton represents about half of the Earth's photosynthetic product, that is to say, the plant itself is the result of the very first and most basic chemical reaction to the sun's rays on sea water. It is born of that chemical reaction, and in a sense turns the sun's energy into an edible form. The smallest animal in the sea, the copepod, lives off the phyto-plankton. As we have seen on p. 65 the copepod has the vital ability to turn much of the phytoplankton it eats into a waxy substance which it stores in a cavity in its body. When it is eaten by predators, and they in turn are eaten by others, the original energy source passes from body to body, distilling and intensifying all the way, until finally, at the end of the chain, it is ingested by man. He is, in a sense, eating a distillation of the sun's light, provided originally, right at the other end of the food chain, by the tiny phytoplankton.

But untreated sewage, industrial waste, rubbish, fertilisers, chemicals, radio-active material, pesticides – all affect the phytoplankton. They either kill it, maim it, or cause it to function wrongly in some way. For 20 years now, scientists have been warning of the steadily diminishing amount of phytoplankton in the world's oceans. If they continue to disappear, then one day the oceans could become vast sterile bodies of water, devoid of life. They would still be inexhaustible sources of mineral wealth, providing our factories with more material than they could ever handle. And we could then dump the enormously increased amounts of industrial waste we would be producing back into the ocean without having to worry about destroying any life. There would be none to destroy.

The choice is ultimately the one we face with every application of technological development. Do we continue to use the ever-increasing number of tools and materials science is putting into our hands, and reap the short-term benefits, with disregard for the long-term effects on us and our environment? Or do we deliberately attempt to slow down and redirect the energy of our scientists to give us time to decide which way we want to go? The choice is ours, but it will not be ours for long.

How to sink 100 tons of crude oil pollution in 15 minutes. A new anti-pollution technique devised by Shell uses sand which has been treated with a chemical that makes oil cling to the individual grains. When sprayed in the form of a slurry, the oil sinks to the bottom — and stays there. A ship this size could carry enough treated sand to dispose of 2500 tons of oil pollution.

4   The Big Clean-up

Earth hath not anything to show more fair :
Dull would he be of soul who could pass by
A sight so touching in its majesty :
This city now doth, like a garment, wear
The beauty of the morning.

The sunrise view from Westminster Bridge in London inspired William Wordsworth's sonnet a full century before the word 'environment' had become associated with a problem. So much was the environment taken for granted that the word was seldom if ever used. Today the world recognises that the preservation of this planet as a place capable of supporting human life is the most pressing problem of our time. In many ways the dangers seem to have replaced the possibility of a nuclear holocaust, which was the dominant nightmare of the late 50s and 60s. But only in the last five years has the public conscience and consciousness been focused in this new direction. That prophets of doom-by-pollution, like our former colleague Gordon Rattray Taylor, have become bestsellers in the United States is indicative of both the cause and effect of this remarkable swing in world-wide popular opinion. There is no lack of concern at all levels from Wigan, Lancs, to Tulsa, Oklahoma. Yet as well as being the greatest single threat to our continued existence, this challenge is unquestionably the most complex which Man has ever been obliged to face.

Among its many paradoxes is one particularly apparent in the British Isles, the experience of which may be relevant to writers of future legislation elsewhere. When the first Clean Air Act was passed by the British Parliament in 1956, it was an important milestone in the history of social legislation. The idea of establishing large urban and industrialised areas as 'smokeless zones' was as courageous as it was original. Now, flying over the British countryside on a clear day, a traveller will be struck by the absence of tell-tale smoke plumes from most industrialised areas. Yet, because of its overall absence, any pockets of air pollution show up like chalk on a blackboard. What is particularly depressing is that a high proportion of readily observable smoke trails comes from plant belonging to nationalised industries. The Electricity Generating Boards, the National Coal Board and the British Steel Corporation appear regularly to violate the spirit, if not the letter, of the Clean Air Act.

For anyone experienced in the ways of Authority, it is not difficult to understand the acceptance of this apparent national scandal. It springs from the natural human reluctance of one British Public Servant to attack another. The power station across the River Thames from St Paul's Cathedral, slap in the middle of Wordsworth's inspirational canvas, is allowed to deluge Wren's masterpiece with untold volumes of smoke while the proceeds from a £1 million appeal to the public's generosity are being spent on cleaning and restoring the corroded stonework of the Cathedral. At Battersea, Fulham, Acton, and all over the Metropolis, indeed from almost every other power station in the country, this state of affairs persists.

Nor is this pollution by Public Authority confined to the air we breathe. The discharge of sewage effluent, manifestly below the minimal requirements of what any right-minded citizen might expect, darkens many once beautiful

Sunrise over the River Thames. And the pollution hangs heavy in the morning air.

waterways, destroys the fish, and reduces the tidal banks to evil-smelling wastes of slime. Protests tend to be met by a statement of a simple and incontrovertible fact – remedial action would cost public money. The views of ratepayers themselves are rarely sought, but it is only fair to point out that the possibility of a majority vote in favour of economy rather than conservation cannot be outruled.

Similar conflicts of values are strikingly relevant in the underdeveloped countries. It is no secret that when, under the mounting pressure of public opinion, a leading chemical manufacturer in the United States announced his intention to discontinue the production of DDT, within 48 hours he was visited by a deputation from India. It was brought to his attention that his well-intentioned policy would without doubt increase the infant mortality in many parts of the world by a factor of at least 50%. Exercising a statesman-like compromise, the manufacturer decided to continue the production of his DDT-based insecticide but changed its name.

It can be argued that a society tends to get the kind of police force it deserves. Today it would seem apparent that Man will get the environment which he deserves – and specifically that which he is prepared to pay for.

An interesting side-effect of this new-found public concern has been to put the blame on technological advance, and accordingly, to discredit it. The cancellation, temporarily, of American plans to build a supersonic civil air transport will go down in history as one of the most extraordinary manifestations of this attitude. It must seem to many people deeply concerned with the problem as a whole that this particular gesture was motivated by emotional or even political considerations rather than a realistic appraisal of the facts. To the more cynical European observer it must seem at best improbable that the powerful American aerospace industry, supporting as it does such a large element of the American tax-paying public, would for long accept the possibility of an economic advantage being allowed to slip into the hands of its Anglo-French and Russian competitors; or that such a situation would be accepted by most Americans.

Be that as it may, the fact that technology had become a dirty word in the United States prompted President Nixon's scientific adviser, Dr Edward E. David Jr, to say in January 1971, 'Recent attacks on science and technology have led to a timidity and hesitancy across a wide range of national activity.' A visit to West and East Coast Universities some months later revealed a hasty rewriting of research programmes to place their emphasis on 'mission-oriented' work, in order to secure the funds for the continuation of research and the existence of entire departments.

At the University of California, Riverside, for example, organised research in even the agricultural and biological sciences is undergoing a more critical self-evaluation than ever before. To quote Dr Lowell Lewis, Associate Dean of Research at UCR's College of Biological and Agricultural Sciences: 'The proposed state budget for the coming fiscal year has less funds earmarked for university organised research than the current one. This alone would necessitate a critical self-evaluation. More important is the fact that we have new avenues of research which we can and must explore, and current projects which demand *more* attention and support if we are to meet society's crying environmental needs and those of agriculture itself.'

One abandoned project is the potential development of geothermal steam fields, discovered by geologists, which might provide desert agriculture with a new large source of usable water. As examples of some of agriculture's pressing problems Dr Lewis cites the need for new pest-control systems that will cause less environmental pollution, and more information on the best use of fertilisers for maximum production and minimum pollution.

Few scientists on either side of the Atlantic would quarrel with that. Recently a distinguished panel hastily convened by the American National Academy of

Science, recommended new Federal mechanisms to consider the broad social consequences of advancing or (significantly) retarding particular technological developments. Its report stated that such developments in the US result from 'a multiplicity of decisions in industry, Government and the market-place'. When individuals, corporations or public agencies considered exploiting or opposing particular developments, they attempted to project the potential gains and losses to themselves, and their decisions usually turned on what they believed to be most in their own interests. This system had created an economy of great strength and versatility, but it also gave rise to 'troublesome inbalances' that were evident in such phenomena as polluted air and deteriorating cities.

The report was nothing other than a published admission that self-interested analyses of the kind usually made in the United States were likely to ignore implications for the rest of society.

Despite the ground-swell of American public opinion against technology, the most positive discernible move to come to grips with the environmental problem in the US has been to establish a Laboratory for Environmental Studies at the Massachusetts Institute of Technology. The Institute is justly proud of its tradition for providing technological solutions to the problems of the Nation. At the head of this new Department is Professor Raymond F. Baddour, until recently Principal of the Department of Chemical Engineering. In his spacious office in one of MIT's maze of new buildings he told us: 'The environmental area is very difficult. When I first agreed to take the job as Director of the Environmental Laboratory, I took what now appears to be a simple-minded attitude. Go to Washington, talk to some people, and find out what are the urgent problems. One of the things I found as I visited different agencies was that they thought it a marvellous idea that MIT should set up a mission-oriented laboratory on environmental problems and they wished us well. One of the major contributions we could make was to help define the problem! That was interesting! Everything was going just fine until I reached the point of talking about supporting work and they said, "Well, gee, we'd love to see you do it but we don't have any money." So at all three levels – the third is implementation – it was going to be difficult. In some of the problems we were considering, many studies had already been carried out. But so far as we could tell, the work might just as well not have been done because the follow-up, the implementation, had not been carried out. For instance, there was no visible change in anything as a result of the studies on whether or not the country should have a national energy policy.'

This is a classic example of complexity. International relations are relevant. A war in the Middle East could completely change the US attitude and policy towards energy. What would be the consequences of an attempt to control the increase in the consumption of energy by, for example, inverting the rate schedule: the more electric power used the more the charge per kilowatt hour? It would certainly affect the Gross National Product of the US and some areas of industry would be hit harder than others. Would the nation stand for it?

There are two competitive processes for making chlorine and caustic, which have many commercial applications. In the last decade the most popular has been one involving the use of mercury cells. But anybody building a new chlorine-caustic plant today in Britain, Canada or the US would not even consider it, despite its many advantages over older methods, because of popular concern about mercury consumption. A similar example is the use of heavy metals in pesticides. Scientists are now aware that it is less important *where* you release an insecticide than in what *form* it is released. If you use mercury compounds to protect the seeds, rainwater extracts the chemical, it gets into the rivers, the oceans, and then spreads all over the world. Professor Baddour is amongst those who think there will be a restriction or total elimination of some heavy metals in certain types of organic compounds.

But he predicts major consequences. Whole industries, and the paper industry is one, have very severe limitations on how much they can do to control the pollution that their plants produce. Yet there are competitors for paper products that can be produced under less polluting circumstances – for instance, synthetic paper from a petroleum-based compound. But this, too, creates a problem: synthetic paper is hard to get rid of once you have made it. So work is going on to find a way to destroy this new plastic by using ultra-violet light to reduce it to powder.

Following scientific anxiety about heavy metals in pesticides is the increasing concern about any pesticides that are persistent 'on a broad gauge', of which DDT is the prime example. All we know about DDT is that it decomposes very slowly. We don't know precisely how slowly, but it has been estimated that over half the DDT that has ever been made is still floating around in the environment. One of the results of concern about persistent and broad-gauge pesticides will be an increasing amount of research into biological rather than chemical methods of control. Releasing large bodies of sterile males to control the population growth of insects is an example. But even that can have consequences that are difficult to predict. You wipe out an undesirable pest, which happens to be the food of a desirable creature, which suffers accordingly, and subsequent pest outbreaks reveal the disaster. Professor Baddour is in distinguished company when he says: 'None of these things can be done without some concern about the side-effects. The really new development is increasing awareness by everyone that we need more carefully to examine the consequences of our invasion of nature's balance.'

A fascinating example of this type of biological pest-control has been achieved by a research team headed by Dr Mir S. Mulla and Dr Toshiaki Ikeshoji at the University of California, Riverside. It has long been known that many species of animals including Man react negatively, even psychotically, to overcrowding. The Riverside research team hopes to use one aspect of this fact to control mosquitoes without resorting to chemicals which might pollute the environment. Their initial research revealed the surprising fact that two specific substances are given off by hemmed-in mosquito larvae subjected to an unnaturally overcrowded environment. These 'overcrowding factors', as the scientists called them, were then put into water where other larvae were growing. The result was the total arrest of larval development. The young mosquitoes either died within a short period or remained within the first or second molting stage, dying a few days later. Five species of mosquitoes proved from 97% to 100% susceptible to the 'overcrowding factors'.

Although other scientists had previously noted that overcrowding retards mosquito larva growth, this was the first time that any of the chemical substances responsible had been isolated and studied. To obtain the materials, Dr Ikeshoji extracted a concentrated residue from water teeming with larvae. Even small doses of the extract proved highly toxic to the larvae. Approximately 95% were killed by the time they reached pupation and the material remained toxic for 2 weeks or more after it was first introduced. Since these chemicals are produced by mosquitoes themselves and are present in nature, their use should cause little or no environmental pollution. And because of the nature of mosquito larval development, it is not necessary to kill all mosquitoes in their larval state in order to control them. Mosquito larvae usually develop in temporary pools, irrigated crops or ditches. The water lasts only a few days in most cases, but this is long enough to permit mosquitoes to reach adulthood. What has to be done, then, is to slow down the larval growth-rate so that the water dries up before the larvae change into pupae and reach adulthood. This kills the mosquito just as effectively as outright extermination, but with much less effort and probably no harmful effects on wild life or other insects.

The toxic 'factors' are given off by crowded larvae, but they are not materially affected themselves. It is the larvae of succeeding generations which are affected. Those that are not killed show abnormal behaviour, slow heartbeat, and a very slow growth rate. Chemical isolation and identification of the overcrowding substances will be the next step.

Another research team, headed by Dr Eldon L. Reeves of the University of California's Division of Biological Control, had made a less complicated but equally surprising discovery in the relentless war against the mosquito: garlic is fatal to them. In laboratory tests garlic wiped out 100% of five species of mosquito including the toughest and most bloodthirsty found in the pasture land of the San Joaquin Valley – aedes nigromaculis. The discovery, which was largely accidental, resulted from Dr Reeves's investigation of certain algae as a possible biochemical control for mosquitoes. He noted that the algae smelled like garlic. The laboratory results of exposing mosquito larvae to minute doses of garlic extract were so convincing that the favourite plant of every French cook may soon become the source of a potent new mosquito larvicide.

In the past many active insecticidal materials have been derived from natural plant sources including nicotines, pyrethrenes and rotenones. Dr Reeves points out that garlic was used in ancient India and China as a folk medicine. It is known to be useful in counteracting worms, protozoa, bacteria, flatulence and urinary troubles. Its active ingredient, allicin, has been shown to be broadly active against various human diseases. Dehydrated garlic has also been found to work against certain causes of food poisoning, especially *salmonella typhimurium*.

A particularly interesting aspect of this research is that although garlic is widely grown, its production could be increased enormously, especially in under-developed countries, for local use against mosquitoes and possibly other insects.

Dr Earl Oatman is another leading figure in the California-based campaign for more efficient pesticides that are less dangerous to the environment. His approach is to employ known parasites as the killer. He has controlled the two-spotted mite on strawberries by distributing amongst them predatory mites imported from Chile. By releasing a tiny parasitic wasp imported from Europe, he achieved a big reduction in the imported cabbage-worm population in local cabbage fields. And with a native parasitic wasp, he controlled the tomato fruit worm which is the larval form of an insect also known as the cotton bell worm and the corn ear worm. It is considered to be the most dangerous agricultural insect pest in the United States.

Dr Oatman distributed his predatory mites following the pruning and mulching of the strawberries. Eight weekly mass releases, at the equivalent rate of 320,000 per acre, effectively suppressed the two-spotted mite pest, which is the most serious pest on strawberries in California. Dr Oatman's parasitic European wasp was able partially to control not only the imported cabbage worm but also another pest known as the cabbage looper. Since there are several hosts for the beneficial insect, its chances of becoming permanently established in Southern California are enhanced. Even so, it is still probable that weekly mass releases of the parasite would be necessary to control the imported cabbage worm on all Californian cruciferous crops.

Dr Oatman considers, however, that his most significant success has been in controlling the tomato fruit worm. After weekly mass releases of the anti-fruit worm wasp in June and July, he found less than 2% infestation in tomatoes at harvest.

It is developments such as these which will be amongst the principal pre-occupations of a new department of the University of California – the International Centre for Biological Control. There they hope that an information-retrieval system can be developed which will allow scientists, in a matter of minutes, to extract from the filing programme information on the natural enemies of any given pest in any area in the world. It would be especially helpful to researchers seeking biological control solutions to pest problems in developing countries.

But it is not only insect pests which are being subjected to the new techniques of bio-control. Two Italian weevils imported by the University and the US Department of Agriculture in 1961 are slowly but inevitably reducing the thorny seeded vine, a particularly persistent weed in Southern California. By 1966 the insects had taken hold and spread throughout the area. No further releases were made.

One type of weevil attacks the stem and crown of the weed, the other infests its fruit and destroys the seed. It is calculated that the weed will not be eradicated but reduced to tolerable levels. The process will take a long time – at least 8 years and probably more – since that is how long the seed can live buried in the soil.

To European gardeners this particular development may not appear to be of dramatic significance, but to pet and bicycle owners of California the 'puncture vine' is well named. It spreads along the ground, distributing its spiky thorns for tyres and paws alike – to say nothing of the naked feet of itinerant hippies or holidaymakers.

Parasitic insects have also been imported from Japan, India and Pakistan in an attack upon an arch-enemy of citrus farmers – the white fly. In their native India and Pakistan the parasites are so effective that the white fly is comparatively unknown. Unfortunately, as well as preying upon the white fly, these insects possess antisocial cannibalistic habits, which has made the importation to California of suitable working colonies a hazardous affair. However, the problems seem to have been overcome and the immigrants are now being supported by two further parasites from Mexico as a means of controlling another recently arrived pest – the woolly white fly.

Scientists and legislators at grips with pollution are aware that the effects of measures to combat it tend to place unevenly distributed loads on various sections of the community. Since air, of all the elements, is common to every environment in which man can survive, the problems of air pollution enjoy the doubtful distinction of increased complexity. The first task has been to establish scientific standards defining what is and what is not acceptable. The complication of such criteria involves gathering all known information on human health and welfare, domestic animals, materials, and a further consideration now gaining increased recognition – the delineation of potential aesthetic criteria. Only after painstaking research for information under these and many other headings can satisfactory air-quality standards be set.

Among the difficulties involved in air-pollution research are the damaging and dangerous effects of a complicated mixture of compounds, when it is difficult to say which of the components may be responsible. Furthermore, the effects of some pollutants may be indirect and only become clinically apparent after many years of exposure. Since the situation is urgent, remedial action should be taken now based upon information already available, even though such information may not be as complete as might be desired.

At a recent symposium there was scepticism about the efficacy of projected US air-pollution control measures, and scientists all over the world are increasingly expressing such doubts. It has been proved, for example, that current regulations for the control of the exhaust gases from motor vehicles are in some ways making a bad situation worse. Toxic particles in exhaust fumes are now being fractioned into such minute proportions that, instead of being washed out of the lower atmosphere by droplets of moisture, they are continuing to drift in ever-growing deadly clouds. Dr Rodney Beard, Professor of Community Medicine at Stanford University Medical School, has cited calculations which show that, even with the proposed exhaust controls required in 1975, the oxidant concentration in the Los Angeles area in 1985 will still be ·2 to ·25 parts per million. The California

Air Resources Board states that deleterious health effects are to be expected with ·1 parts per million for one hour.

An alternative approach has been suggested by scientists working in the field of energy generation. By using amounts of power by no means excessive in relation to that consumed hourly in present-day highly developed urban communities, it would be entirely feasible to ventilate artificially whole areas affected by smog. The Los Angeles basin is the classic example. But although there must be very few city dwellers in the Western world who are prepared to deny the undesirability of smog, much remains to be ascertained about its true effects.

A minor component in smog is ethylene. It is a growth regulator produced by plants in quantities sufficient for their own needs. In small quantities ethylene can stimulate growth and cause buds to open. It is used to speed the ripening of citrus fruits and bananas. But ethylene is also a significant ingredient of automotive exhaust gases. It has now been discovered that in concentrations such as those experienced near motorways and dense traffic concentrations, it is strong enough to harm nearly all plants. Research has clearly demonstrated that tomatoes, almonds, roses, orchids, snapdragons and carnations are all sensitive to ethylene and injured by it. Other gases such as ozone and peroxyacetyl nitrate are well known as growth inhibitors affecting many crops from citrus to spinach.

It is only very recently that ethylene has been shown for the first time to be a component of smog capable of really dramatic growth-inhibiting effects. Professional carnation growers in the Monterey area of California were puzzled by peculiarities in plant growth and loss of blossom production, and they suspected the effects of air pollution. A detailed study undertaken at their request revealed that carnation plants exposed to even very low concentrations of ethylene grew only half as big as similar plants growing in clean air. Stems were shorter, and flowers smaller.

It remains to be seen whether the carnation industry along the South Coast of France will be adversely affected by the construction of the new Riviera motorways. There is as yet no known way to protect plants against ethylene injury, but anti-oxidant sprays may some day offer a method of protection against smog damage. Nothing is yet commercially available, and the best advice research botanists can offer is for growers to select varieties and strains of plant resistant to air pollutants. Some varieties of tobacco plants, pine trees, tomatoes and lettuce have been shown to have strong smog-resistant capabilities – cold comfort to the carnation growers.

On the other hand, it has been proved that the lead particles discharged in the exhaust of motor traffic are *not* causing soil pollution to the extent anticipated by some experts. Although Dr D. Page, a soil scientist working in the University of California, has found that the concentrations of lead in soil surfaces in the Los Angeles area, where traffic density is high, has increased by two or three times over the past 40 years, the current concentrations are not considered excessive. It has already been established that lead in the atmosphere near major highways ranges between 5 and 15 micrograms per cubic metre – about one tenth of the maximum level prescribed by industrial safety demands.

During his study of the lead content of crops grown near major highways, Dr Page found that the amounts of lead in and on unwashed tissue of plants grown less than 66 ft from the roadside were from two to five times greater than those found in and on crops growing 500 ft or more away. Ordinary washing can remove from 50-90% of the lead, depending upon the type of crop. The higher concentrations were due to airborne particles landing on the plants rather than by absorption through the soil.

Conventional methods of horticulture – breeding young plants from seeds, cuttings or graftings – are no longer sufficiently precise for the demands of modern plant technology. At an International Horticulture Congress in Tel Aviv Dr Toshio Murashige has reported an alternative approach. He has two main objections to the traditional methods. First, plants developing from seed are often exasperatingly variable; if a lot of money has been invested in valuable plants, the economic risk of producing unmarketable variants can be too great. Secondly, plants propagated by conventional methods are subject to disease organisms that can systematically infect and destroy them. Dr Murashige's alternative, from which he believes may spring many important economic and ecological developments, is to use tissue culture as a way of propagating plants.

So far Dr Murashige and his team have grown more than fifty kinds of plant, from asparagus to rare trees, in racks of test-tubes under pink lights in a controlled laboratory environment. The first step is a detailed understanding of the plant from which tissue cultures are to be produced, including the precise food, light and temperatures required.

Under germ-free conditions almost invisible pieces of tissue are excised and cultured in test-tubes. So far many kinds of plants have been successfully reproduced, including citrus varieties, daisies, petunias, sweet williams, lilies, gloxinia, potted foliage plants, ground cover plants, ferns and gerbera. Current work is concerned with oak, palm, conifer and various flowering shrubs.

Dr Murashige predicts that the tissue-culture technique will have real economic impact on commercial fruit, flower and vegetable farming once improved crop quality and expanded production are achieved. But the development of strains free from disease is probably an even more exciting possibility. The magnificent Temple Orange, which is subject to crippling virus disease, could become a commercial crop because its inherent virus can now be eliminated. The Neyer Lemon, a notorious carrier of viruses and a popular garden tree in citrus-growing areas, could be guaranteed virus-free if grown from culture tissue.

The technique also holds particular promise for growers of asparagus. When grown from seed asparagus varies greatly from the parent stock. Growers would benefit considerably from being able to grow asparagus which duplicated a superior mother plant.

Inevitably the comparison of plant-tissue culture to animal-tissue culture comes to mind. If a single cell or group of cells can be made to grow into an entire plant from the test-tube, what about animals and people? Dr Murashige points out that while he has enough information on the behaviour of plant cells to cause them to develop into a root, a stem, a leaf or a flower, his knowledge of animal-cell behaviour is insufficient. He plans to confine his work to the field of plant-tissue culture in which, as he says, 'The job to be done is quite enough for any one research group.'

For those who believe that despite the difficulties Man is still capable of overcoming his environmental problems, there is one unfailing source of hope and inspiration – the passionate and active interest in the subject on the part of young people, particularly students. Nor is their interest confined to meetings of protest, demonstrations, and study groups. A voluntary ban of the use on their campus of non-returnable plastic containers and packaging has been imposed by the students of at least one university in the United States. The target of their self-discipline is, of course, the very real hazard of pollution by plastic litter.

There are two essentials for any successful campaign against pollution – public awareness and practical alternatives to the pollutant. Both requirements were admirably fulfilled by the Clean Air Car Race of 1970. It all started in 1968 with a coast-to-coast race between an electric car from MIT and one from California Institute of Technology. The cars were built and driven, and the race organised, by students. It was as a result of that race that two members of the MIT team, David Saar of New York and William Carson from Wisconsin, issued a challenge in October 1969 which resulted in the departure of 42 cars at dawn on 24 August 1970 on a 3600-mile, 7-day, cross-country 'dash' from Cambridge, Massachusetts, to Pasadena, California.

The Saar-Carson car used a petrol piston engine to provide intermittent power for an advanced solid-state alternator of new design. The alternator in turn recharged the car's bank of conventional batteries which supplied an electric motor that actually powered the vehicle. The piston engine emitted low-pollution exhaust because it ran at a constant speed. During city driving it was turned off and the car ran on batteries alone. Another entrant in the race was a turbine electric pick-up truck with a 600-hp engine driven by a kerosene-fuelled turbine generator mounted in the back. A third MIT car, a steamer, was still in its garage late that afternoon, although the students building it retained hopes of getting it on the road eventually.

When the race began, altogether 28 American and two Canadian colleges and two American high schools were represented. The 42 cars that took to the roads that morning must be one of the most technically interesting collections ever assembled. They included five electrics, two electric hybrids, two steamers, one turbine, one diesel, and a variety of modified internal combustion engines, either burning unusual fuels, or using exhaust emission-control systems, or both. The internal combustion cars included thirteen operating on liquid propane gas, six using natural gas in any one of several forms, six that burnt unleaded petrol, two using leaded petrol, one alcohol, one alternating between propane and natural gas, and two that burnt mixtures of two fuels – methanol/leaded petrol and liquid natural gas/hydrogen.

The most spectacular start by all accounts was made by the red pick-up truck powered by a 1000-lb electric motor and a paraffin-burning turbine generator. The turbine was giving a fair imitation of a jet engine at take-off in the enthusiastic hands of its 22-year-old MIT student driver, urged on by his Faculty Adviser, Professor Charles Draper, whose laboratory designed the guidance system that took the Apollo astronauts to the Moon and back. The aim of this particular entry was to show that heavy-duty electric motors are a solution to the problem of reliability in electric propulsion. The noise factor was generally agreed to be as yet unsolved!

The University of Toronto was the target for the first over-night stop, and subsequent rest halts were scheduled at Ann Arbor, Michigan; Champaign, Illinois; Oklahoma City, Oklahoma; Odessa, Texas; and Tucson, Arizona. The winner's flag awaiting them in Pasadena the following Sunday must have seemed to many of the starters a good deal more distant than the Moon.

The back-up organisation was also impressive. Counting 'trail vehicles' of various descriptions, the cavalcade for clean air included an 11,000-gallon liquid natural gas tanker. Charging stations had been set up every 40 to 70 miles along the route for the benefit of the electric vehicles. Propane dealers had been organised into a refuelling network. Many unleaded petrol cars were carrying spare drums of their particular fuel, and more drums had been cached in areas where such fuel was no longer commercially available.

The event was perhaps more a rally than a race. Not the least of the formidable obstacles which confronted the organisers was to work out a system for deciding who had won. It was agreed that there should be a winner in each of the five classes (modified internal combustion, electric, electric hybrid, steam, and turbine) and an overall winner. But here any similarity with conventional motor sport ceased. Basically the vehicles were competing against pollution; at the same time they had to be seen to perform well enough to be acceptable to the general public. Unfortunately, reducing pollution emissions from the exhaust generally lowers the quality of a vehicle's performance.

The scoring system was therefore reduced to an equation designed to take

account of performance, exhaust emission and reliability: $S = E(P + R + TE)$, where S = total race score; E = a pollution emission factor; $P \& R$ = performance and elapsed race time scores; and TE = thermal efficiency score.

The 'P' factor was derived from a special test in which points were won for acceleration, braking, noise level, and an autocross. 'R' represented the total number of points gained on a conventional rally basis on each stage of the race (scheduled time against actual time of arrival at controls). 'TE' was really a pure economy rating. Because of the many different types of fuel used in the race, a miles per gallon figure would have been meaningless. Instead all types of fuel – petrol, propane, natural gas, electricity, etc. – were converted into energy units in the form of British Thermal Units, and the thermal efficiency of each vehicle judged in miles per BTU.

'E' was the most important factor in the scoring. Each entrant could more than double his performance and elapsed time scores by lowering his pollution emission efficiently. 'E' was the ratio of actual emission to that specified in the 1975 Federal Standards for exhaust levels. Thus an entrant whose car exactly met the 1975 standards would have an E score of 1. Vehicles with emissions better than that would score 1 plus. If an entrant reached the 1980 Federal Standards, which are considerably more stringent than those planned for 1975, he received a bonus of $0 \cdot 3$ in his emission factor, bringing it up to $2 \cdot 5$. Curiously, however, any entrant who managed to better the 1980 standards still only netted an E of $2 \cdot 5$.

Enthusiastic organisers of British motor sport will be amused to learn that student race officials on the road had a computer with which to calculate daily scores. Using teletypewriter consoles in the mobile headquarters-van, and in the

National Race Information Centre in Chicago, they had access via private telephone lines to a time-sharing computer in Minneapolis. But so great was the diversity of performance between classes that the organisers took the responsibility for picking an overall winner out of the hands of the computer and gave it to a panel of five leading figures in automobile technology and exhaust pollution.

The elected overall winner was a 1971 Ford Capri powered by an internal combustion engine equipped with fuel injection, after-burner, an exhaust-gas-re-circulator, and four catalytic mufflers attached to the exhaust. The car was fuelled by lead-free petrol and designed by students from Wayne University in Detroit.

In the internal combustion engine class (fuel in gaseous form) the winner came from Worcester Polytechnic Institute – a 1971 Chevrolet Nova modified to run on propane. In the fuel-in-liquid-form category of the same class, Stanford University won with a 1970 American Motors Gremlin running on alcohol. The hybrid electric class result was a tie between Worcester Polytechnic Institute and the University of Toronto.

The turbine electric vehicle from MIT was the only turbine entered and therefore came first despite arguments about faulty emission tests, which made the compilation of its total score impossible. Finally, there was no winner in the steam class because all the entrants dropped out shortly after the start.

Each winning team received a grant of $5000 from the National Air Pollution Control Administration, a Federal Agency which has also undertaken further tests on each of the vehicles to assess their potential.

The amount of publicity generated in the United States was disappointing to the organisers. In Europe the event passed virtually unnoticed, perhaps due to the influence of major advertisers with a vested interest in the *status quo* of the automotive industry. But there is to be another Clean Air Race in the United States in 1972.

The Clean Air Race was, in a way, further proof that all over the world the love-affair between man and his motor-car seems to be coming to an end. Even the 'in-laws' of this relationship, the manufacturers, are suffering from the effects of this disenchantment – not yet perhaps in economic terms, but certainly in terms of attitude. When Ralph Nader took on General Motors and won, he was the self-appointed mouthpiece of a fast-growing body of opinion. And when on a visit to Britain in the spring of 1971 Henry Ford said that the quality of products from his British factories left much to be desired, despite the outcry that followed he was only expressing what a very large proportion of car owners know to be true. Nor may that criticism be confined to the products of a single firm or factory.

The root cause of general discontent is probably the result of environmental, as much as technical, considerations. The motorist is fed up with spending hours of frustration on roads no longer capable of meeting the demands made on them. He is tired of breathing polluted air and uneasy in the knowledge that he himself is adding to that particular hazard. And he is seriously concerned about the safety of himself, his family and his neighbours as a result of the present chaotic and deteriorating state of road traffic, and the resultant accident figures.

In our last book we reported on the moving-passenger conveyor, developed by the Batelle Institute and others, as being the most promising solution to the threat to cities of death by traffic strangulation. There remains the equally pressing problem of long-distance, city-to-city road traffic which, despite the investment of millions of pounds and the sacrifice of important environmental considerations, has not by any means been solved by the construction of motorways.

It must by now be abundantly clear that we can no longer afford ourselves the luxury of transport wholly under our personal control. The average driver is incapable of meeting the demands made upon him in many traffic situations, and any attempt to improve the 'throughput capability' of highways, while leaving the onus of ultimate responsibility on the individual driver, can expect only a very limited future. But so well established is the convenience and pleasure provided by the family car that any attempt in the West to deny it to the average breadwinner is politically difficult, if not impossible. The responsibility of contemporary technology must therefore be the provision of an acceptable alternative. A great deal of work in this field has been done at research institutes in Europe and the United States. The consensus of informed opinion seems to point inescapably to some form of guidance system which will relieve the individual driver of responsibility for at least major sections of any cross-country journey.

Most of the solutions being developed or proposed are in the group that has come to be known as 'flow systems'. They are designed, like escalators, to operate on a non-stop, minimum-wait basis to counteract the most unattractive feature of traditional public transport – the combination of time involved in waiting, getting aboard, and finding a seat, and the generally low average speed resulting from frequent stops. A basic disadvantage of non-automated transport is that there has to be a comparatively large space separating individual vehicles or trains, which necessitates large-capacity vehicles. A train carrying perhaps 500 passengers is therefore obliged to slow down and stop from 100 miles an hour for the convenience of a handful of passengers who have reached their destination.

By definition a 'flow system' is one in which vehicles keep moving on a track or guideway, and move off before slowing down to stop. A motorway is a flow system in which vehicles operate under manual control. The addition of automatic control makes possible a much higher density of vehicles. One track or lane of almost all automatic control flow systems proposed at present will, at full speed, accommodate between four and ten times the number of vehicles. So striking are these advantages that it can be safely assumed that the next generation of overland transport will use an automated flow system of one kind or another. It is only in the area of choice between systems that technologists differ. In fact there is such a diversity of alternatives that engineers are apt to confuse not only planners and politicians, but the public as well. This abundance of ideas is in itself a good thing, but the time for decision is near. It is therefore important that the public, who are going to have to pay, should be made as fully aware as possible of what such a decision will mean.

On the evidence available it seems that the most promising solution is an automated palleted system. This appears to offer the best hope of providing low-speed urban systems as well as inter-urban passenger and freight services of up to 300-mile stages. The palleted system is one in which a normal road car is driven on to a carrying unit – the car itself thereafter remaining stationary. The present-day inter-city, car-carrying train services are a form of palleted non-automated flow system.

Palleted Automated Transport (PAT), which has attracted the attention of many engineers, including Professor David Wilson and his team in the Mechanical Engineering Department of MIT, is a more advanced development of the same idea. It is designed to accommodate unmodified passenger cars, minibuses, and freight containers. Perhaps surprisingly, the MIT team chose an unflanged, unpowered and unbraked steel wheel for their load-carrying pallets. Propulsion is by vertical axis synchronous electric motors which drive shaft-mounted cog gears on each pallet. The cogs engage a stationary rack gear mounted on the left-vertical guiderail of the guideway. This is therefore a vertical, rather than horizontal, configuration of the rack and pinion mountain railways of a century ago. The problem of points or switches which bedevils any system employing flanged wheels or monorails is overcome by movable guidance arms on both sides of the pallet which can engage either the right- or left-hand guiderail on the permanent way.

With PAT the problem of accelerating to the controlled speed of main flow traffic has several alternative solutions. The synchronous motor could be supplied with a varying frequency current, a magnetic linear accelerator could be used, or a series of worm drives could be applied. Since this last offers the most precise methods of locking-on to the main rack and pinion, it must be a strong contender.

As long ago as 1950 General Motors and RCA developed a method for applying automatic control to private cars by giving them pick-up aerials and devices which permitted operation of the accelerator, brakes and steering from signal coils embedded below the road surface. A similar device was demonstrated at the British Road Research Laboratory in May 1971 and, although it offers specific improvements on the original work, it is significant that General Motors and RCA did not proceed 20 years ago because of what they described as 'unresolved concerns' about the reliability of the system, and questions of legal liability in the event of an accident. The serviceability of each individual car and its additional equipment is also fundamental to the safety of such a system.

Several other dual-mode systems are at present under discussion or development. (A dual-mode system is one in which the same vehicle can be operated under the direct control of the driver, or locked on to an automated flow system.) These systems, such as pallets, have specific advantages. The main-line parts of the route involve substantially less land acquisition than a motorway. The ability to go in one's own car from home to destination at high average speed and with few waits makes this substantially more attractive than either existing urban services or many of the new rapid-transit ideas. Although dual-mode systems with private cars will probably not be acceptable for city centres, bus-type vehicles can be used for delivery at destination and can be intermingled with through vehicles on the same guideway. It is this kind of integration which makes the pallet system so attractive. Other advantages are:

1   A pallet system taking standard vehicles has a much lower initial cost than one in which specialised vehicles have to be used. For the same system utilisation the number of pallets needed would be approximately 1% of the number of special vehicles.
2   The reliability of pallets owned by the operator of the system would, because of design, maintenance and inspection-standards, be higher than that of modified private cars. To achieve equivalent reliability, each private car would have to undergo very extensive checking every time it entered an automatic guidance system employing its own power and controls. This would involve a severe penalty in time and cost.
3   Pricing and taxing policies could encourage the use of low-powered, perhaps battery-driven cars.
4   A palleted system is totally suitable for the automatic distribution of containerised freight.
5   The pallet system has distinct advantages in noise, fumes, land-acquisition and landscaping.

On the other hand, pallet systems have the disadvantage of requiring empty pallets to be circulated in order to anticipate demand, and some storage space for them has to be provided at every entrance. Even so, it is thought that the cost of this could be absorbed more readily than in other systems.

Given the practicality of a comparatively simple rack and pinion drive, despite the advantages of monorails, magnetic or air hover, and other recent advances, the choice of unflanged steel wheels for the pallets is an interesting example of the way in which many engineers are today deliberately avoiding the newest techniques in favour of the simplicity, reliability and cheapness of more traditional methods and materials. Because directional control is achieved by a guidance rail at the side, there is no necessity for flanged wheels. They run on a resilient surface of hard rubber or plastic, thereby reducing noise to a minimum. The necessity for high wheel-to-track adhesion is obviated by the rack and pinion drive. This enables maximum advantage to be taken of the position control of an automated flow system, and minimal spacing can be maintained between pallets even at high cruising speed. The use of synchronous electric motors in conjunction with the rack and pinion completes this simple formula for precise positioning.

As yet MIT's PAT consists only of a 1/20th-scale working model, but were the sum total of work at many European and American centres to be co-ordinated, a full-scale operation could probably be launched in a few years.

The fact is that unless we are capable of bringing every available technological resource to bear, we are unlikely to solve our major environmental problems in the limited time available.

It is in pursuit of further knowledge about the potential use of computers in the control of the environment that a unique experiment has been conducted by students of the Architecture Machine Group, a Ford Foundation research effort within the MIT Urban Systems Laboratory. The principal subjects in this extraordinary drama are a small colony of gerbils, and a machine called SEEK – a sensing/effecting device controlled by a small general-purpose computer. It is a mechanism that senses the physical environment, directly affects it, and in turn attempts to handle local unexpected events within it. It is, in fact, an intelligent building machine which can stack, align and sort 480 toy blocks. But SEEK's endless problem is that these blocks form the environment for a colony of gerbils, and the gerbils busily push and bump into the blocks, topple towers, and disrupt the painstaking construction of the machine. The result is a substantial mis-match between reality and the programme instructions retained in the computer's memory. SEEK's role is to deal with these inconsistencies and in so doing it exhibits inklings of responsible behaviour – the gerbils constantly disrupt their planned environment and the reaction of SEEK is purposefully to correct, or alternatively amplify, these gerbil-provoked dislocations.

The world which SEEK and the gerbils share is an open-topped, transparent-sided tray measuring 5 × 8 ft. A miniature travelling crane, guided by the computer, is capable of selecting and moving individual blocks. By contemporary standards of computer techniques, this arrangement is trivial in its simplicity. Yet SEEK metaphorically goes beyond the real world in which machines cannot respond to the unpredictable nature of people. As yet machines are poor at coping with unexpected changes in what they have been led to expect. SEEK's purpose is to overcome this lack of adaptability, because if computers are to be our friends, they must learn to understand us. If they are to be responsive to changing, unpredictable, context-dependent human needs, they will need an artificial intelligence that can cope with complicated contingencies in a competent manner, just as SEEK deals with elementary uncertainties in a comparatively simple-minded fashion.

SEEK's software (which gives it its instructions) puts its hardware (the electronics of the travelling crane) into one of six modes of operation: Generate; Degenerate; Fix it; Straighten; Find; Detect Error. Each mode deals in its own way with the 2-inch blocks spread out across the whole area of the tray in a pattern that does not match the computer's model, because as the machine places the blocks the gerbils move them.

The Generate Mode involves a random number generator which has been programmed to build enclosures out of the blocks. It is impossible to predict precisely where Generate Mode will place the blocks, but it is certain that they will be built into nooks, crannies and mazes in which the gerbils play happily, disturbing and rearranging as they go. Nevertheless SEEK determinedly continues arranging its blocks until all 480 have at least been placed as intended.

The Degenerate Mode occurs when all the blocks have been used and SEEK is forced to take them from one part of the site in order to stack them in another. This process involves long pauses for 'thought' (from 30 to 60 seconds) as SEEK goes through more than a million computations necessary to determine the appropriateness of removing any given block.

These two Modes largely determine the form of the gerbils' environment so far as the machine is concerned. The remaining 80% of the software capability handles the unexpected events – either 'gerbil-provoked dislocations' or machine malfunctions.

It is the 'Fix It Mode', as for most of us the machine's most difficult challenge, which occupies it for most of the time, and involves more calculations than any other. A single disruption – a block at the wrong angle, one missing or one too many – is good evidence that other similar dislocations may exist. Each of these apparent dislocations has to be diagnosed, since any block can be in one of three conditions: it can be slightly out of line from its original position and SEEK must realign it with other adjacent blocks; it might be an inch or so out of place, and therefore have to be moved; or it might be a relatively long way from its original position as the result of a gerbil action.

Straighten Mode is necessary to transport a crooked block for realignment – so that it will stand with its rectangular sides parallel to the boundaries of the container tray. The preceding modes call for Straighten Mode when their own operation is hampered by a misaligned block. It is picked up and carried to the block straightener (a plastic box to one side) which is designed to reorientate blocks dropped into it. If the force of the fall alone fails to align the block, the computer turns on a vibrator. As soon as it is straight, and depending upon which Mode has called upon the services of the straightener, the block is either returned to its original location or a pile of blocks at the far end of the pen. Find Mode is also used by the other modes of operation. Since SEEK does not have an eye, it must use this mode to 'feel around' for the blocks. Finally, in the Detect Error Mode the machine checks itself for both hardware and software malfunctions, and when one is detected it warns its operators (not to say the gerbils) with a siren and flashing lights.

Needless to say, SEEK has been a resounding success with human visitors of all ages, and the gerbils give every indication of enjoying their relationship with a machine.

Although the SEEK experiment is relevant to the application of computers to many environmental problems and the development of more refined machinery for the purpose, it is no accident that it was conceived by students of architecture. Of the countless constituent parts which make up our environment, perhaps the most significant is the houses in which we live. Another project at MIT has been to explore the potential of a computer for designing custom-built homes in urban areas. In contrast with the suburbanite, the urban dweller's plot of land is actually a three-dimensional space, which is not necessarily on the ground but lodged within a framework of high-density accommodation in a block or tower. The goal is to make possible the attainment of a particular user's needs and desires in physical form, and at the same time to foster in him a better understanding of his needs and the consequent architectural problems involved.

It is well known that one of the principal difficulties in the architect/client relationship is one of language. Most people are unaccustomed to communicating graphically, and are probably not very good at expressing their ideas in drawings anyway. They prefer to resort to verbal description which may be disastrously inadequate. For this reason much of the software in the project is devoted to the solution of this problem, and accepts the responsibility for teaching as well as acquiring information.

It is anticipated that a great deal of the initial time that a given user will spend with the machine will be devoted to learning how to express himself in graphic terms. As soon as possible in the dialogue between the user and the computer-architect the machine will encourage the user to draw, and will comment on the results. The prime function of the machine is to learn about the user, since its knowledge of architecture will already have been programmed into it. The machine will build a model of the user's new or modified home, but it will simultaneously be building a model of the user and of the user's model of it. As the dialogue progresses and the machine gets to know the user, its questions will become better, more relevant, more incisive, and more revealing.

The Architecture Machine argues for a more directed effort towards machine intelligence, because computer-aided design demands understanding of context and meaning in the real world. But a further important goal is to help people to analyse and express their own domestic needs and desires, and to understand more fully their own patterns of living and how they affect and are affected by the physical environment in which they live.

Projects of this kind, though largely theoretical, suggest exciting possibilities for the next generation of householders. Meanwhile work goes on to try to cope with the needs of today's generation.

Many of the tasks as yet scarcely delineated as 'desirable projects' at MIT and elsewhere in the United States have been the subject of fruitful research for years by several British institutions. Legislation already written by the British Parliament has put into practical effect the products of such research and discoveries. One example is the work of the Warren Spring Laboratory at Stevenage, Hertfordshire, a research establishment in the Department of Trade and Industry. Its work is mainly devoted to the development of the industrial aspects of science and technology. This is concentrated in three main fields: mineral research and technology; chemical and process engineering; and air pollution. Large elements of the first two are what would be described in the United States as 'environmental-related'. The work on air pollution is guided by the recommendations of the Clean Air Council and the ultimate object is to achieve cleaner air in the urban areas of the United Kingdom. It is therefore of service to the country as a whole rather than to particular branches of industry, and provides technical information to the various government departments and local authorities concerned in the administration of clean-air legislation, and to make an assessment of its effectiveness.

With the co-operation of local authorities, the survey of air pollution by smoke and sulphur dioxide involves daily measurements at some 1200 sampling sites in over 350 towns and cities, and in rural areas throughout the United Kingdom. This survey is designed to assess progress towards clean air, to discover where the need for remedial action is most urgent, and to provide information for studies of the effects of pollution on health. Knowledge is obtained in every type of district and community. Warren Spring Laboratory is already providing in Britain precisely what has been asked of the newly established Environmental Laboratory at MIT for the United States, that is to 'define the problem'.

The British survey is based on simple equipment which is owned and operated by local authorities, has in many instances been developed by the laboratory itself, and which can be used by relatively unskilled people to make the necessary daily measurements. The laboratory supervises the work, publishes the results, and is responsible for interpretation of the data. The type of apparatus used has been accepted by the Organisation for Economic Co-operation and Development for use throughout Europe wherever similar measurements are required. A modified form of the apparatus enables daily samples to be taken even though the observation site can only be visited once a week. The scale of the survey varies from whole towns to particular streets and sometimes even various points in a single building.

Curiously enough, there is no systematic air-pollution survey of London, but sufficient measurements are made at over 170 sites by local authorities and others in the Greater London area to give a picture of the pattern of pollution by smoke and sulphur dioxide in the metropolis. A map of sulphur dioxide emission for the London area is being made from measurements taken at various heights from the Post Office Tower. During London's now rare days of smog, the laboratory undertakes detailed surveys in the city to assess the extent and severity of the situation and provide a basis for possible counter-measures.

Measurements of smoke and sulphur dioxide are being made by specialised techniques at a number of rural sites, in order to establish background levels in different areas, and eventually to assess the drift of pollution from town into country. Coastal sites have been included so that, when steady onshore winds prevail, the measurements should indicate the amounts of pollution reaching the British Isles from overseas. Other countries in Europe are engaged in similar studies and the accumulated observations should be of value in assessing long-term changes in regional and global patterns of pollution.

5   The Driving Force

Cape Cod, Massachusetts, is a delightful peninsula of woods and coastal heathland, the favourite of the very rich, the comfortably retired, and a sprinkling of artists and writers. Pleasant, widely-spaced houses nestle beneath trees, and well-tended gardens sweep down to the water's edge. As a result of the Cape Cod community's new-found horror of smoke pollution, bonfires for the destruction of garden rubbish have been banned by statute. If a limb falls from a tree on to one of those artificially nurtured lawns, the wretched householder has to drag it as best he may to the civic tree-shredding machine, or alternatively pay for a similar, but mobile, device to be brought to his property. Unwanted roadside tree-growth is similarly disposed of – and the resultant wood chips may be bought by the local gardeners to sprinkle about the roots of their shrubberies.

Quite apart from the risk of spreading every kind of botanical disease – particularly those caused by fungi – from the widespread distribution of the shredded remains of infected timber, this policy could be fairly regarded as sheer lunacy. To anyone seriously concerned with the major contemporary problems of conservation, such gross abuse of energy-producing resources is nothing short of outrageous.

To give the good residents of Cape Cod their due, some have protested about the local power station which pours from its ugly chimneys a haze of toxic exhaust gases into the once clear air of one of the many local beauty spots. But the power station is to be enlarged because of the increasing demands made upon it – including the considerable requirements of the civic timber-shredding machines.

Like Fred Streeter we too have that 'old-fashioned' liking for the smell of wood smoke, and believe in the efficiency and efficacy – as well as the romantic attraction – of the garden bonfire. But the seemingly bewildered conscience of the Cape Cod community is by no means confined exclusively to Massachusetts.

The problems of power generation, and conservation, are now accepted in scientific circles as being fundamental to the survival of Man (see 'The Big Clean-up'). The dwindling resources of on-shore fields of oil and natural gas have led in large measure to the current emphasis upon off-shore exploration, and the necessary development of new submarine techniques. Anyone reading a newspaper, listening to the radio or watching television will be warned, once a week on average, that the Earth's supply of known mineral energy-resources will be exhausted within thirty years. At the same time, newly discovered fields of immense potential make just as frequent headlines. It is difficult for the average consumer to equate one statement with the other. But regardless of the precise number of years, it is apparent that the ever-increasing demand for more power will eventually, and in the not too distant future, call for more fuel than is available with contemporary techniques. Yet although there must be very few people in authority who are prepared to deny the existence of this alarming situation, it is difficult to discern positive action by Governments or industry towards meeting it. The atomic energy authorities of Britain and the United States, for example, have to struggle for survival against every form of vested interest from the mining unions to the treasury. Fortunately for us all, major figures in the currently much-criticised world of science and technology have been for some years at realistic grips with the problem, and have advanced the only solution to the dilemma so far discernible.

The complicated mass of pipes and equipment sitting on deck is Yves Cousteau's diving saucer, the SP 350. The entrance is through a watertight hatch (centre). Recently Cousteau used the minute Alsthom fuel cell to power all the equipment on board — including the battery of lights used for filming (bottom centre). (See p. 106.)

In our previous book we drew attention to two comparatively recent developments which are of particular relevance – the exploitation of super-conductors, and research into nuclear fusion. These are at the spearhead of the technological counter-attack upon the problems of power supply. But lower down the scale there is a wide perspective of valuable research taking place in laboratories all over the world.

It has long been known that if you mix hydrogen peroxide and hydrazine in the right proportions, under the right conditions, and in the presence of a suitable catalyst, you get oxygen, nitrogen and electricity. The problem has been to control the reaction in such a way that the electricity could be put to practical use. In a laboratory a few miles out of Paris, the scientific research team of a comparatively small French company set about overcoming the difficulties about 9 years ago. They have now produced a practical, working fuel cell which so excited the distinguished French oceanologist Commander Jacques Cousteau that he immediately ordered four for installation in his newest submarine research vessels, and instructed the company to build a larger version for delivery as soon as possible.

Submarine research (see p. 70 ff) is only one of many requirements which have been unsatisfied by the efficiency of conventional electric batteries. The power-to-weight ratio offered by even the most elaborate and expensive types of modern battery is too poor to be acceptable for contemporary demands. The electric-powered car is a case in point: batteries weigh too much for what they can give, they run down too quickly, and they take a long time to charge up again. Research into various types of fuel cell has been conducted energetically in recent years, not least by the major car manufacturers of Europe and the United States, backed by the enormous resources and facilities which only they can command. But at present it seems probable that this small French team may have shown them all the way to a practical solution.

In order to control the nature and timing of the reaction between hydrogen peroxide and hydrazine precisely, a special synthetic semi-permeable diaphragm, or membrane, has been developed. It allows the two substances to intermingle in minute quantities through its microscopic pores, and the result is a phenomenon known as aeonic transfer. The next task was to design a container which made it possible to introduce the two substances in controlled conditions on either side of the diaphragm. Electrodes were made by plating the faces of non-porous plastic wafers with a catalyst appropriate to either substance. Each individual fuel cell consists of a 3-decker sandwich – the plastic-backed electrode and its catalyst; the semi-permeable diaphragm; and the next and opposite (+ or −) similarly-made electrode. The total thickness of this sandwich is *less than one millimetre*.

The inlet and exhaust 'pipes' are to a similar scale, and consist of a series of micro-grooves etched into the plastic at the top and bottom of each of the electrode plates. A similar series of micro-grooves etched in a herring-bone pattern (the result of painstaking research) ensures the even and controlled flow of the liquids as, separated and intermingled by the diaphragm, they are drawn from the bottom to the top of each cell. An inert liquid is used as an electrolyte to 'carry' the appropriate quantities of hydrogen peroxide and hydrazine to the cells.

Each of these 1-mm, 3-decker sandwich cells is capable of generating one volt. Join 240 together in a pile of 'sandwiches' mounted upon locating pins, and you have a cell complex capable of producing 240 volts, yet only 24 centimetres high.

The remainder of the device is merely plumbing, albeit as ingenious and elegant as the construction of the cells themselves. A pair of rotary pumps, operated by the electricity generated by the cells, draws the two separate flows of hydrogen peroxide plus electrolyte, and hydrozine plus electrolyte, into the cell pack, across the faces of the electrodes, and out through the exhaust 'pipes'. The 'exhaust' fluid is then passed to a cooling chamber, followed by a gas separator

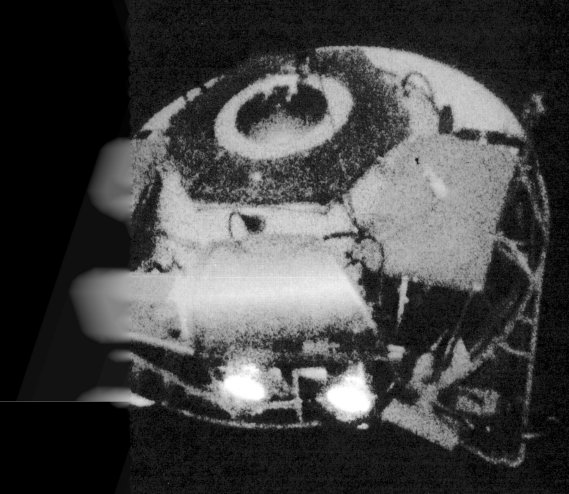

which draws off the unwanted oxygen and nitrogen. (This could of course be simply released or collected and used for other purposes.) The electrolyte is then re-energised by the addition of further supplies of hydrazine and hydrogen peroxide which are drawn from their separate containers and mixed into the supply-stream through a venturi nozzle which ensures the correct quantity.

The device is what is known as 'autonomous' in that it will continue to produce electricity on demand, automatically running its own pumps at the correct speed for re-energising its electrolyte as long as the supplies of hydrogen peroxide and hydrazine are maintained. The efficiency of this French fuel cell can be gauged in terms of its power/weight ratio which is 6-10 times better than conventional batteries. Its potential applications are enormous, particularly in situations demanding simplicity, reliability, and a minimum space/weight penalty. And that, as any engineer would confirm, is a considerable range of requirements.

The production of oxygen in the French power cell makes it of particular value in submarine research or any other situation in which a breathable but artificially created atmosphere must be produced. NASA has of course dedicated an immense research effort in this direction. The maximum exploitation and conservation of energy and life-giving materials is of paramount importance to the survival of any space capsule.

But there are many curious paradoxes in the uses made of the enormous amounts of power generated by conventional aircraft. As yet, the weight/space penalty of what have become traditional techniques has not invalidated them from application to even the most advanced passenger-carrying airliners in contemporary service. But the advent of Concorde and its Russian counterpart has postulated more exacting demands to be met by the designer. As an example,

the comparatively inefficient way in which an emergency oxygen supply is ensured in contemporary airliners is unlikely to be imitated by the operators of Concorde.

Prototype 5 kW Alsthom fuel cell which runs on hydrazine and hydrogen peroxide. (See p. 106.)

The recent outbreaks of hijacking have raised the question of what happens to the passengers within an aircraft flying at over 15,000 ft should a bullet or grenade puncture the skin of the pressurised cabin; and apart from such extreme emergencies, the possibility of the mechanical failure of an airliner's pressurisation system has always had to be taken into account. Experienced airline travellers will be familiar with the repetitive charade performed for their benefit by the cabin staff. The automatic way in which an oxygen mask drops from above the passengers' heads is controlled by a simple barometric device which responds to any excessive drop in the atmospheric pressure within the cabin. Oxygen is supplied to each mask as the passenger pulls it towards his face; the pressure is calculated to facilitate effortless breathing and sufficient oxygen to compensate for the decrease in cabin pressure. What the airline operators refrain from pointing out – and who can blame them? – is that in order to meet this emergency requirement, they are obliged to build into their aircraft apparatus which is not only expensive in terms of space and weight, but a major potential danger in case of fire.

Each mask is fed by a ring-main system which has to run twice the length of the aircraft fuselage. In the new generation of jumbo jets and air buses this means some 300 ft of tortuous pipeline containing free-flowing oxygen under pressure. Apart from the necessity for thorough and frequent checks for leaks, the consequences of a fractured pipe, highly probable in a runway accident at high speed (which still constitutes a considerable hazard in airline operations), are self-evident. A transatlantic jet airliner carries some 15,000 litres of oxygen under high pressure in several containers; their proximity to burning fuel, following what might have been a comparatively minor accident, is too horrifying to contemplate. There is also the additional need for large quantities of oxygen to be stored at refuelling points – all this necessitated by the requirement for an oxygen supply to be used in emergencies only.

Just as the chemically generated electricity of the French power cell provides a solution to the inadequacy of conventional batteries, a practical alternative to these elaborate requirements would be the development of a chemical oxygen

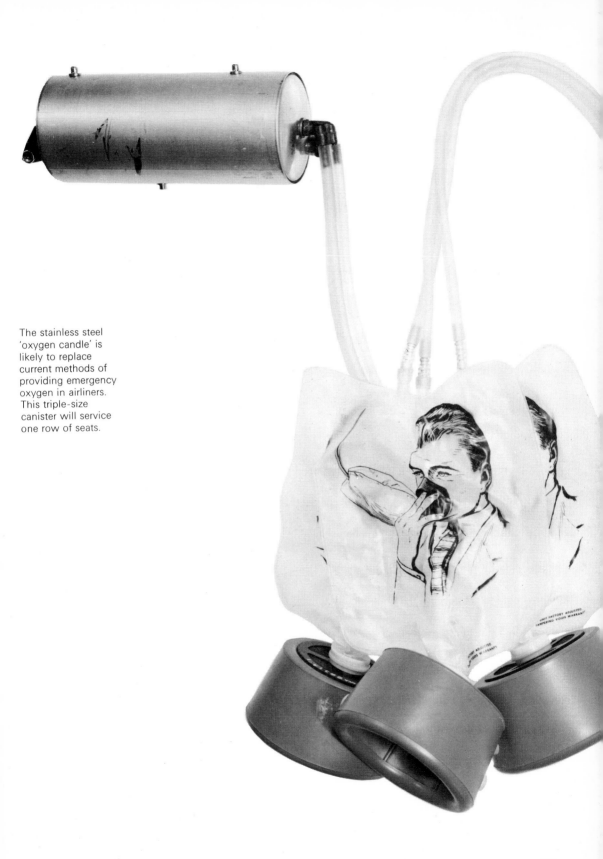

The stainless steel 'oxygen candle' is likely to replace current methods of providing emergency oxygen in airliners. This triple-size canister will service one row of seats.

generator. Such a device, known as an 'oxygen candle', has recently been introduced to the aircraft industry. The candle itself consists principally of sodium chlorate. When burnt it gives out pure oxygen. Contained in a stainless-steel cylinder and 'triggered' by a simple, self-contained ignition mechanism, the oxygen candle can be connected to a conventional oxygen mask in the usual way, and can generate sufficient oxygen for a 15-minute emergency. A barometric device, similar to that currently used, would release the oxygen mask over every seat should the need arise. The action of the passenger in pulling the mask to his face would trigger the mechanism which lights the candle. The real and peripheral advantages of such a system appear overwhelming. It saves weight, space, 'plumbing' and maintenance aboard aircraft, and by localising the potential supply of 'free' oxygen to each candle improves the safety factor in case of fire. Concorde and the European airbus, as well as the Lockheed Tristar and DC10, will probably be equipped with this system.

The objective to which the complex and expensive technology of the modern aero engine should be directed is the high-speed transportation of large numbers of people and heavy loads of freight. But the non-payloads of peripheral equipment carried by airlines have tended to reach alarming proportions in recent years. This reduces the performance-potential of the power generated by their superb engines, undermines their efficiency and adds materially to the cost of air travel. So any contribution towards the decrease of this 'dead weight' must be of major importance.

In fact it seems most likely that it will be economic pressures rather than thoughtful responsibility which will induce an attitude towards the conservation of power more befitting 20th-century man than that which pertains at present. In Britain the switch from coal to North Sea gas has been a political issue of some importance. Nevertheless the sacrifice of the vestigial remains of Britain's once-great coal industry was an economic necessity which no responsible government could ignore. Meanwhile the utility companies and corporations do their best to induce the public to buy ever-increasing quantities of their products, be they electricity, gas, oil or solid fuel. Each group competitively develops new devices to attract more custom, and therefore more demand, regardless of whether the fuel concerned may be the most efficient for the task. For example, the use of electricity for domestic heating may prove, from the point of view of energy conservation, a totally misguided policy. Electricity, in the view of many technologists, should be generated in order to provide mechanical power rather than heat. They go on to argue that the goal of contemporary research should be increased efficiency, and therefore lower consumption for a given requirement. The current policy is merely to stimulate demand in order to increase revenue, an essentially unscientific and highly inefficient situation.

The case for increased efficiency must win the approval of the housewife since it would tend to decrease her electricity and gas bills – or at least make some contribution against this particular aspect of inflation. Unfortunately she is seldom given the chance to make the objective choice which would express her point of view. There are, of course, exceptions, and one has recently been developed by a small private company in Britain whose business is domestic heating by natural gas. Taking as their model the heat-exchanger which kept the Apollo communications system cool, they have developed a device which is now undergoing small-scale field trials in several urban areas. Its purpose is simply to get more heat into the domestic water supply per cubic foot of gas consumed.

The system consists of a perforated metal cylinder into which a suitable gas/fresh air mixture is pumped and burned. The flame propagation is confined to the outer circuit of the cylinder, which is in turn surrounded by a coil of water pipe as in the conventional steam boiler. But in order to maximise the transfer of heat from the burning gas to the water circulating through the pipes, the space surrounding the coil is occupied by a large number of steel pellets the size of duck shot. The steel balls, heated by the burning gas passing through them, greatly improve the efficiency of the heat transfer to the water contained in the pipes. The manufacturers claim an efficiency of 82% heat transfer against an average of 70% attainable in other conventional domestic systems. A simple method of demonstrating its efficiency is to put one's hand into the flow of exhaust gas from the vent pipe. The 18% of waste heat makes the exhaust gas just bearable to the touch – but even so, this simple test underlines the enormous waste which still occurs even in such a comparatively efficient arrangement. Although the average householder may not be particularly concerned about the possible effects of the ever-increasing discharge of heat into the atmosphere, the saving of 12% on the cost of his hot water is likely to win his sympathy.

On a far more dramatic scale it is the overall improvement in the efficiency of energy transmission which is attracting a major research effort into the development of Super Conductors. The physical property known as high-temperature high-field super conductivity will probably have a revolutionary effect throughout the electrical industry within the next two decades. In this context 'high' temperature means approximately 20° Kelvin – about the boiling point of liquid hydrogen. Even this is a higher temperature than has been applied to any practical super conductor so far developed, and current scientific opinion is that it is unrealistic to envisage a super conductor that will work at room temperature.

Super conductors were first discovered in the university of Leiden, in Holland, during the early part of the century. A very dramatic demonstration of their capabilities was made at a meeting of the Royal Society in London in the 20s. A super conductor 'loop' was flown from the Leiden laboratory to London, bearing in it an electric field and current which had been initiated before it left and were still observably active after its arrival. For many years super conductivity was regarded as an intellectual curiosity, and more work was devoted to the discovery of its cause than to the exploitation of its potential effect. It was probably Bernt Matthias and his colleagues in the Bell laboratories in the United States who, more than anyone else, first raised the possibility that super conductivity might be of vital importance in the technology of tomorrow's society.

Matthias pointed out that the operation which our technology accomplishes least efficiently is producing and maintaining a magnetic field. A magnetic field represents stored energy, and it doesn't necessarily require any energy supply to maintain it. But the production of magnetic fields is essential to the use of electrical power because the forces that are exerted on electrical conductors are forces which are responsible for useful work (as opposed to the creation of thermal energy, by electricity) and are always produced by the interaction of magnetic fields. Inefficiencies encountered are the result of resistivity – the resistance that occurs in electrical conductors. The generation of heat as the result of a short circuit in a domestic power supply is probably the most commonplace example.

The waste of electrical energy through the inefficient production of magnetic fields is of immense importance in contemporary technology. As much as 50% of the electrical power used for purposes other than heating is effectively wasted by the generation of unwanted heat in conductors. The super conductor offers a way in which this wastage may be avoided, and consequently it is difficult to over-estimate the potential effect of their widespread application. In terms of the transmission of electrical power, distance would become no object. The concept

A silent diesel engine. It is covered with a special sandwich of plastic and steel which absorbs virtually 50% of vibration and sound. Most of the insulation is clustered around the very noisy parts of the engine – the cylinders, rocker boxes, timing case and valve chest cover. The first prototype completed two years ago is now in the Science Museum, London.

of constructing super-power stations in comparatively remote areas becomes much more attractive if significant amounts of power so generated can be made available in distant cities and industrial areas with minimum loss.

An experimental power line using super conductors, developed by the Edison Electrical Institute and the Tennessee Valley Authority in the United States, is now in its third year (see *Tomorrow's World*, 1970, p. 212). Its purpose is to put to practical test extra-high-voltage power cables, at extremely low temperatures, employing super conductors for practical underground power transmissions. Meantime, all over the world, and particularly in the United States and Britain, the search for suitable materials continues. A super-conductor, when cooled within a few degrees of absolute zero, loses all its electrical resistance. The laboratory search has involved combinations of all the 103 elements that make up matter, and the extremes of temperature, both high and low, to which the materials are subjected has necessitated the development of a technology specific to the purpose. The very rapid production of temperatures approaching 4000°C is required to fuse such substances as conuminium, germanium and niobium into suitable alloys; they are then supercooled to temperatures of around minus 250°C.

In order to meet extravagant demands such as these a special type of experimental crucible has been developed at the Clarendon Laboratory of Oxford University. It has neither base, nor solid sides. The metal to be melted is 'levitated' above a magnetic field generated in the base of the crucible. In the form of a small pellet it hovers, supported only by magnetic energy. In fact four-fifths of the power used in this apparatus creates the magnetic field; the other fifth heats the metal pellet by induction. The interior of the crucible is made up from copper segments. As the metal to be heated floats above them, there is no possibility of metallic contamination, and as a further precaution a water-cooling system keeps the segments at comparatively low temperatures.

While such elegant apparatus may be adequate for the exacting demands of the research laboratory for which it was designed, the production of sufficient quantities of super-conductive material for the construction of any large-scale transmission programme comparable to the National Grid is clearly a very different proposition. The combination of niobium tin and plastic tape developed by the Plessey Company in Britain has the advantage of comparative simplicity and flexibility in all senses of the word, but the advantages offered by super conductors remains a challenge to the metallurgists as well as the physicists.

Any scientist must be amused by the irony of the current advertising campaign of business corporations in the United States which have a vested interest in nuclear-fission electricity-generating plants. In the glossy pages of the more expensive magazines and trade journals, they attempt to set themselves up as 'the non-polluting power source', completely ignoring the fact that the by-products of nuclear-fission generating stations are amongst the most deadly materials known. The disposal of radioactive waste is already presenting a worldwide problem, and the accidental discharge of immense quantities of radioactive material in lethal concentrations remains an undeniable danger. This is not to say that the risk is unacceptable within the confines of the strict measures of control and emergency procedures under which nuclear-energy electricity stations are built and operated. Nevertheless the ethics, and indeed the morality, of a publicity campaign deliberately designed to promote a public image which pays no regard to these considerations must be at best questionable.

There is an alternative to nuclear fission reactors, with their dependence for fuel upon uranium or its man-made derivative, plutonium. It uses fuel derived from the cheapest and most readily available of all sources – water; its waste product is an inert and harmless gas – helium. This is the nuclear fusion reactor –

Five new ways to get about

Magnetic train *(top left)*. This German prototype model literally floats on air. It is held in suspension by the force of magnets embedded in the rail. Forward propulsion is provided by a linear motor. It is virtually silent and runs at about 300 mph.

Kindrue steamer *(middle left)* — a two-seater sports car powered by steam. Behind the boiler are condensers which turn the exhaust steam back into water to give greater mileage. It has rear-wheel drive; no clutch or gearbox are required. Cruising speed: 50 mph. Top speed: 60 mph.

Fast Cat *(bottom left)*— this version can reach 60 mph. The next model should touch a hundred. Stability at these speeds comes from twin catamaran hulls and the steeply V-ed keels.

Top right — Experts predict that turbines like this one will be common on the rounds within five years. Their main advantage over internal combustion engines is that they are smaller, more powerful, and produce less pollution.

Inset — British Leyland's latest gas turbine lorry. It is still at the prototype stage.

British Rail's Advanced Passenger Train *(bottom right)*. Has a top speed of 150 mph and cruises at 125 mph. It should be in experimental service by 1975.

the control and harnessing of the energy-producing process of the sun itself. Like the stars, the sun is a huge nuclear reactor, but unlike the nuclear reactors already built by man its process is one of fusion as opposed to fission. Nuclear fission produces energy by splitting the nuclei of heavy atoms; nuclear fusion is precisely the opposite – it liberates energy by joining together the nuclei of light atoms. But the problems of building a practical fusion reactor for the generation of useful amounts of electricity are enormous.

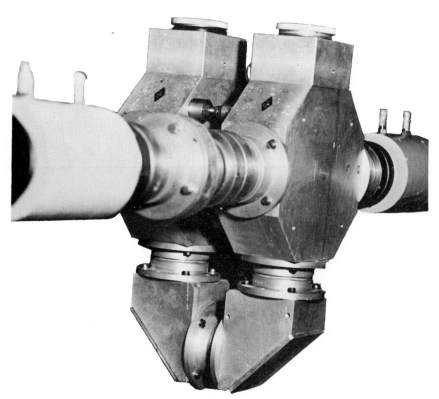

(Above) Circular Waveguide German Post Office engineers developing this telecommunications system plan to get it in service by 1985. The idea is that instead of wire cables, low-frequency radio waves will be beamed along tubes like these, called waveguides. The radio waves behave like light so they have to be reflected around corners. The advantage of this system is that more individual signals can be packed in a tube than in a conventional telephone cable of the same diameter.

Within this wave-guide device is the highly polished prismatic systems (left) that bends low-frequency radio waves around tight corners.

Although fuel for a nuclear fusion reactor could be derived from water, it is in fact tuterium – the sort of hydrogen that makes 'heavy water'. In nature tuterium is extremely rare – only one atom of hydrogen in every 10,000. However, in the form of plasma the energy-potential of tuterium is so great that the nuclear energy which could be derived from a gallon of sea water would be about fifty times the chemical energy released from burning a gallon of petrol.

To be effective such a plasma would have to be raised to a temperature of 20 million degrees. The problem is then how to contain it in a vessel. There are two ways in which this might be achieved. One is to hold the gas together for such a very short time that, in the words of Dr Bill Thompson of the University of California at La Jolla, 'it didn't even know it was there'! This is known as inertial confinement, and has been dramatically demonstrated as feasible in such places as Christmas Island in experiments associated with the testing of nuclear weapons.

The other method is to hold the gas in place by electric and magnetic forces, and in Britain, the United States and Russia work in this field has been going on for many years. But even this can be likened to the task of carrying water in a sieve. Magnetic fields are almost infinitely flexible and elastic, and the gas exerts enormous pressure; it is also extremely mobile and can therefore get through the holes in the 'sieve', no matter how painstakingly designed and constructed it may be. Demonstrating that it could be done was the principal achievement of the great Zeta experiment at Harwell in 1957 just before the second Geneva Conference on the peaceful uses of atomic energy.

Zeta was a large, doughnut-shaped vacuum chamber in which a very large electric current was carried by a plasma. The current produced its own magnetic field, so shaped that it confined the plasma that was producing it. Early observa-

TV by waveguide. At the receiving end of the hollow waveguide pipe a special filter detects individual low-frequency radio waves and turns them into electric signals that can be fed into a conventional cable system. Waveguides will be employed when video telephones come into common use. They are more economical than any tele-communications system developed so far.

tions at Harwell, however, demonstrated that if nothing more was done the resulting current, in a sense disorientated from the centre of the vessel, would behave in an extremely unstable manner and follow a confused and intricate pattern. To be kept in the centre of the vacuum vessel it had to be 'stiffened' by the influence of an extra magnetic field which was so generated as to encircle the containing doughnut at right angles to the field produced by the plasma's current.

The worldwide publicity for this achievement created a situation with which British technologists (and the British public) are now all too familiar: too many hopes were raised too high. Now after more than a decade of further research all over the world, Zeta is recognised in scientific circles to have been a very important experiment indeed. Gradually the results of measurements and observations made at the time are being more widely understood and recognised, and it is undeniable that the Zeta experiment has had a profound influence on the subsequent development of worldwide research in this field.

The Zeta experiment and similar achievements demonstrated that there was in Britain one of the world's most vital and important laboratories in the field of nuclear physics. Perhaps there still is. Twenty years ago, when he was with the British Atomic Energy Authority, Bill Thompson was one of the founders of the Culham Laboratories. Working at Harwell at the time he says that he was rather opposed to its foundation because it put a rather far out kind of research in a position where the Treasury could attack it too obviously. Dr Thompson considers that subsequent events have proved his fears to have been well justified. His opinion is shared by others of his colleagues at the time, including Professor Bill Drummond who is now working in the University of Texas. They believe that the Treasury systematically made the wrong judgement in every possible case, including that of the Culham Laboratories. 'It is an absolute tragedy to see the damage done to the morale in the laboratory by the announced cuts in funds. There are a number of dedicated people who are still working there despite this discouragement, but a number of the very best people have left.'

Since the Zeta experiment leadership in this field has passed from Britain, the United States and Germany to Russia, which has been extremely active in the field of theoretical physics. At the time of the Zeta experiments, the Russians were propounding various advanced theories which were well understood at Harwell. At the Kartov Institute in Moscow a device was built which succeeded in confining plasma in a stable state for long periods. But it is important to understand the time scale involved in experiments of this kind. The time taken for an unrestricted particle to escape from a plasma container is something like a millionth of a second. The Russians managed to confine their particles for about one hundredth of a second, which means effectively that the particles were attempting to escape from the system 10,000 times before they succeeded.

Two years ago, again at the Kartov Institute in Moscow, Professor Zamokiv produced an extraordinary device called a Tokomak, which differed from Zeta in one fundamental respect. In Zeta the principal current ran in the plasma itself; in the new device it ran in external windings. The success of the Russian machine was remarkable: it confined a plasma at a temperature of 10 million degrees for periods in the order of one tenth of a second. A little-publicised triumph in international scientific co-operation resulted in the temperature measurements of the Russian experiment being made by a team from Culham, who confirmed the estimates that had been previously made by the Russians. About seven Tokomaks with various minor modifications are now being built in the United States.

Bill Thompson, like scientists all over the world, is unabashed in his praise of the Russian achievement. 'In a way it is a little embarrassing that this should have happened, because we have a great scientific potential in the United

States and have advantages which the Russians lack. But the Russians have a sort of optimism which I sometimes feel we lack. They really believe in the future of man, and sometimes I feel that many of us have lost faith in the world of technology and its ability to cope with the problems of society.'

So far science has succeeded in producing plasmas which are at thermonuclear temperatures and which have been confined for approximately the right amount of time to produce significant amounts of energy. But there is a third factor. The densities of the plasma so confined are still too low for appreciable amounts of energy to have been produced. Density is critical to the amount of energy because it results from the collisions between pairs of particles: the number of such collisions depends first of all upon the number of pairs of atoms, and that depends upon the square of the mass of the plasma.

The scientific feasibility of a thermonuclear fusion reactor capable of feeding power into a national grid adequately has not yet been demonstrated. Even then it will have to be proven that such a device can be made into a practical engineering project. A pilot plant would have to be built. It is conceivable that a scaled-up Tokomak would work, but because there is still a certain vagueness about precisely what happens inside one, no one – not even the Russians – is prepared to say that, if one were made ten times bigger, it would produce the necessary amounts of power. Judged by the time scale of some other if less important technological achievements, progress is slow. There are two reasons: scientists working in this field are anxious to make sure of the ground as they go ahead; and, so far as Britain and the United States are concerned, funds are not being made available for carrying on the research at an adequate level. The total American budget in this field is only $25 million. One American utility company alone is run on a budget of about $10 billion a year. In Bill Thompson's words, 'Our children will have to pay for this'.

But if, as we believe, the future of energy supply has assumed crisis proportions, any assessment of potential solutions must be considered on the basis of some sort of time scale. Apart from Tokomak there is another type of nuclear fusion device capable of producing usable amounts of power which is being studied. This is a 'pinch' device, a modified version of Zeta, the Culham team had intended to build in that laboratory before the cut in government finance for the programme. Within the next 5 to 6 years the experiments yet to be completed with these two devices should establish the scientific feasibility of thermonuclear fusion as a potential source of usable energy. However, the history of work in this field has revealed that, even given the best possible theoretical analysis, research teams can still encounter the unexpected which may delay, or even speed up, progress.

Blockage Buster. The power for this gun comes from a CO_2 cartridge. A squirt is guaranteed to unblock most sinks, lavatories, and drains within two minutes.

After the satisfactory completion of feasibility studies, as Dr Thompson has pointed out, the problem becomes one of engineering. And it is at this point that the current work in the field of super conductors becomes vitally relevant. Thermonuclear fusion appears a much more attractive proposition today than it did ten years ago because of the existence of high-field super conductors and their promise of providing, at a comparatively low price, the magnetic field that is necessary for plasma confinement.

The next probability is that the first use of a thermonuclear reactor may be not as an independent power source, but as something complementary to a fission reactor. It could be employed as a substitute for a fast breeder reactor because of its ability to produce a very copious flux of neutrons. Even though in the early stages it may not be possible to build a thermonuclear fusion device that will produce a sufficient surplus of power over and above its own requirements to make it a practical proposition, it might provide a source of small neutrons cheaply enough to compete effectively with the fast breeder. If this proves to be the case, its development as an alternative would be urged by those concerned about the safety aspects of fast-breeder operations. At the present time, the fission reactor has been proven as a practical source of power, and the breeder reactor, thanks to the dramatic achievements of its courageous devotees, has been demonstrated as feasible in scientific *and* engineering terms. It is quite possible that the future of the world's power supply could be assured with the fission reactor and fast breeder as a method of producing fuel. But without such breeders it seems that fission energy has a comparatively limited future.

There remains an alternative source of energy supply to which surprisingly little attention is being paid in Britain and the United States, but much more elsewhere, including Russia and France. If one 10,000th part of the solar energy reaching Earth could be employed as true, rather than thermal, energy, it has been calculated that this would be sufficient to support the industrial requirements of contemporary technology and its anticipated demands to the turn of the century.

One possible step in this direction may be closely associated with attempts to establish a permanent space station in orbit. A suitable reflector could be installed on such a device in 'stationary' orbit so that it could beam solar energy towards Earth as radio power, rather than thermal energy. This would obviate the losses encountered by absorption in the Earth's atmosphere, the possibility of distorting the thermal balance of Earth, and the possibility of major effects on our climates.

An even more advanced approach might be more attractive to future generations of physicists than the current lack of enthusiasm might suggest. At present small groups of people are studying the dynamics of the Earth's atmosphere and the oceans in order to win energy from them. But these studies raise the question of what limits must be applied to Man's use of power, with its consequent release of excess heat into the atmosphere.

Meantime, in France, a programme of practical pilot plants making direct use of the sun's thermal energy has been in operation for some time. Situated near Montlouis in the Eastern Pyrenees, the project continues the experimental work started in Algiers before the break-up of the French colonial empire. There a solar furnace generating 40 kilowatts was operated. A 50-kilowatt plant was subsequently built at Montlouis itself. But now a giant concave mirror, concentrating 1000 kilowatts at its focus, has been built at a nearby site at Odeillo, 5000 ft up. A system of mirrors covering an entire hillside collects the solar energy, but even so, the average of 175 sunny days in the year does not compare with the 300 days available in the Sahara on the original site. However, in view of the experimental nature of the work, it was decided that the sacrifice would be more than compensated by the proximity of technical resources available within the frontiers of France. The solar energy experimental complex is part of the National Scientific Research Centre (CNRS) which has been established to conduct general research into solar energy, to explore the problems associated with its capture, and to develop feasible applications.

The problems of concentrating solar rays of sufficient energy on a specific point, such as a crucible, are considerable. The use of lenses is impractical, so a concave parabolic mirror is employed. The principle is the reverse of that in which a car's headlight is focused to produce a parallel beam from a single light source. Since the sun must be followed in its path across the sky, the reflective apparatus has to be able to move. One solution would be to make the whole mirror pivot – this was the method adopted in the Algiers solar furnace. But it is difficult to mount a large parabolic mirror capable of smooth and continuous rotation, and the suspension of objects in its focus presents further problems. The alternative is to use a flat-surfaced mirror to follow the sun; turning slowly on a mechanical device it reflects the sun's rays constantly into the major parabolic mirror, which remains stationary. In fact this is quite simply achieved if the flat mirror is rotated about a north-south axis at half the speed of the rotation of the Earth. This is the system that has been adopted at Odeillo, the only difference being that, since it would have required an enormous pivoting mirror to feed sufficient energy to its parabolic counterpart, 63 separate flat mirrors have been erected on the hillside, each directed to the single paraboloid.

Each of these moving mirrors turns automatically towards the sun like so many robots attracted to the light. In front of each mirror is an optical piloting device which consists of a lens and a flat matrix of photo-electric cells. So long as the image of the sun, focused by the lens, is formed in the centre of the photo-electric cell plate, nothing moves. But as soon as the image projects over any off-centre cell, a signal operates a system of small hydraulic rams and levers which rotate and tilt the mirror as required. At sunrise the impression created by this small army of automata dotted across the hillside, as they slowly pivot to greet the first rays of the sun, is positively uncanny.

The huge concentrating paraboloid has a total reflecting surface of 2000 square metres, or five times the area of the Centre Court at Wimbledon. Surfaces as large as this are of necessity made up of a considerable number of flat or slightly concave mirrors, each of them individually positioned in such a way that the whole forms a paraboloid.

The mirror so formed occupies the entire north face of the main building of the laboratories – north, of course, because the sun's rays are in the first place reflected onto the parabola by the 63 swivelling mirrors. The other three faces of the building are occupied by conventional laboratories and offices, but its northern aspect has the appearance of a huge Aztec temple. A specially constructed satellite building resembling an airfield control tower contains the experimental apparatus at the focal point of the mirror, the beam passing into it via a special window.

At the smaller Montlouis furnace, with a parabolic mirror area of a mere 91 square metres, temperatures of 3400°C are obtained at the focus. This is the temperature of an electric arc, and what used to be called 'raw earth', such as zircon and thoria, can be melted and synthetic precious stones related to rubies and sapphires are produced. Gems such as these are used in the generation of laser beams. Comparatively large quantities of materials can be handled at Montlouis, including the melting of up to half a ton of steel at a time.

At Odeillo, with its parabola more than twenty times larger, temperatures of approximately 4000°C are generated. For smelting and processing at these high temperatures rapidly revolving crucibles are used for certain experiments. The

H

heat beam enters them through the centre, while the material in the crucible is held against its walls by centrifugal force. Quantities of up to 3 tons can be melted in this way with maximum thermal efficiency. A process that appears to have a great future is that of 'calcilating' or agglomerating at very high temperatures. A great advantage of the solar furnace is that it provides one of the most powerful and satisfactory techniques for high-temperature processing without contamination. The use of the conventional crucible can be avoided, there is no risk of contamination, foreign bodies can be excluded even when dealing with large masses. In order to protect the material to be melted, it is sometimes covered with a sheet of cello-

Solar mirror at Montlouis, France.
(See page 120)

phane. Curiously enough the tremendously powerful thermal beam will pass through such a transparent partition without damaging it, so that when a block of steel is melted in this way, the cellophane cover is not destroyed until the steel begins to run.

Apart from its value as a research tool, the solar furnace at Odeillo can meet the industrial requirement for giant crystals, free from internal stresses, which are used in advanced electronics and atomic applications.

It is by no means beyond the bounds of possibility that giant solar furnaces like this one may play a key role in providing materials and techniques which will one day enable us to harness the ultimate in energy-providing resources – the dynamics of the solar system itself.

6   The Inner World

Nearly every woman knows the horror of looking at her own face in a concave magnifying 'make-up' mirror. Pores take on the appearance of Moon craters. A spot can look like Mount Etna at the height of an eruption. Magnified only two or three times larger than life, a laugh-line at the corner of a pretty mouth becomes a fissure that only major surgery could repair. The lady with an inquiring mind might try to use the information from her make-up mirror to predict what she will look like when she is old and frail and seventy. When she plays this game she is embarking on the kind of journey that scientists make when they try to deduce from observed facts the undiscovered vistas that lie just beyond their present knowledge. Look closely at anything and, at the very least, you may learn a little more. Yet very often a closer view does little other than raise fresh questions and new problems. But as any scientist will tell you, that is what research is all about.

Largely as a result of the explosion in technical know-how over the past decade, the tools that make it possible to look closely at the things around us have improved beyond the wildest dreams of the early researchers. It is not so much that higher magnifications are possible today, though that is important. What is significant is the clarity of pictures – the improved definition – that can be obtained almost as a matter of course. With the latest scanning electron microscopes, you may venture into the very heart of a minute structure to gaze at things that no human eye has ever seen before. The beautiful and delicate internal structures of animal cells, the sinister shapes that are cancer, the incredible nature of little-understood substances – all these and much more are within reach using the new tools and techniques of the microscopist's trade. Of course, as with everything, there are potential pitfalls to be taken into account. Some researchers are beginning to believe that it is possible to end up with so much visual and theoretical detail that they risk not seeing the wood for the trees.

In cancer research, for instance, the criticism is sometimes made that too much effort has been devoted to examining the internal mysteries of diseased cells, while ignoring the larger organisms of which the cells are only a tiny part. Yet, as other researchers look deeper and closer, there is a counter-argument which holds that question-begging pseudo-explanations for physical phenomena can be undermined and dispelled only by examining and understanding in detail the component parts of the whole body. There is probably truth in both points of view. The problem is striking the correct balance. And it could just be that one day a researcher will find out who is correct by looking down a microscope and drawing the appropriate conclusions.

Eyes of a male housefly. Each segment of the multi-faceted eye is a lens. The longer lashes are highly sensitive. (mag. x310.)

Cells of the cervix covered by a cancer growth. The use of scanning electron microscopes to aid diagnosis is still in the early experimental stages. The idea is to compare what is already known about the disease against what can be seen in pictures like this to try to set up a diagnostic base-line of information. (mag. x3,500.)

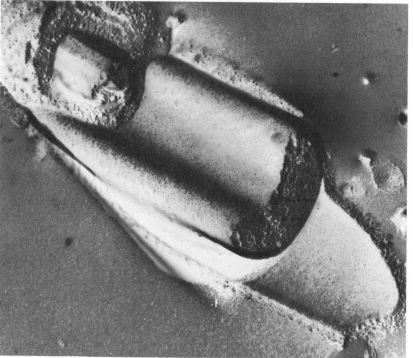

Part of the sperm from a *Roman Snail* (helix pomatia). This photograph was taken on an electron microscope. The roman snail is the one most commonly served in restaurants — cooked with garlic and butter. (mag. x44,300.)

(facing) Live sperm of a *marine snail* (archidoris pseudo-argus). (mag. x2,330.)

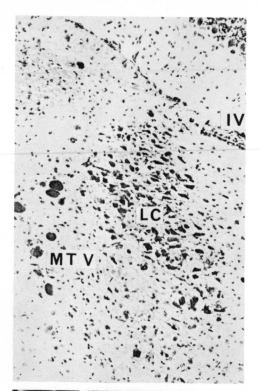

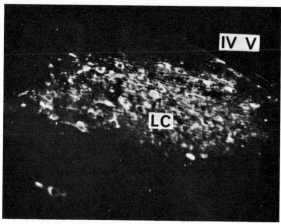

In certain parts of the human brain a chemical called Noradrenaline acts as a sort of postman, carrying information from one nerve to the next. Experiments so far suggest that the information concerns emotional reactions — probably to taste. The picture on the left, taken on a conventional optical microscope, shows the cells that carry the Noradrenaline. On the right, the cells have been stained in a special way to reveal the Noradrenaline. It shows up as a white blob in the nucleus of each cell.

(below) Red corpuscles in monkey's blood. There are about 6 million of these per cubic millimetre of blood. (mag. x1,260.)

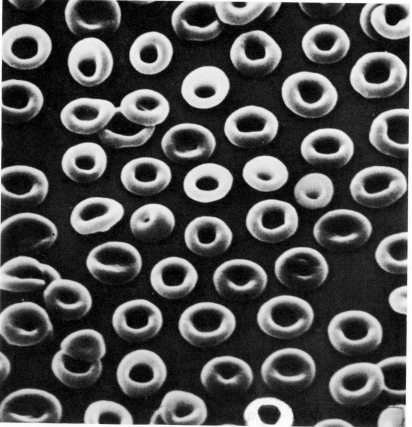

Moon rock *(facing)* These specimens were all brought back by the crew of Apollo 12. The main conclusion geologists have drawn from them is that Earth and Moon were formed from basically the same materials. But there is no suggestion that the Moon was originally part of the Earth. 1, 2, and 3 are estimated to be 3·7 billion years old. 1 and 2 are viewed through a polarising microscope, which shows the minerals pyroxene (orange and blue), feldspar (white and grey), and oxidide minerals (black). 3 shows the same minerals but in their natural colours. 4 is a sample of lunar soil that has been welded together after millions of years of bombardment by micrometeorites. It contains glassy spheres, formed at about 3000°C, feldspar, and other rock fragments. (magnifications: x147.)

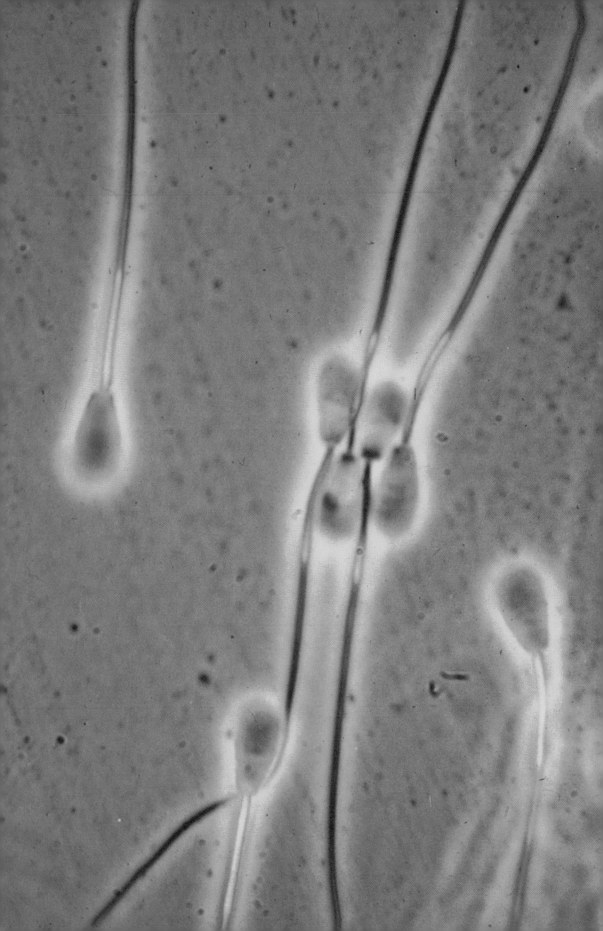

Left: Living bull sperm, viewed by phase-contrast illumination on
an optical microscope.
Top: Living squid sperm (loligo opalescens).
Bottom: Normal living human sperm viewed by interference
contrast microscopy.

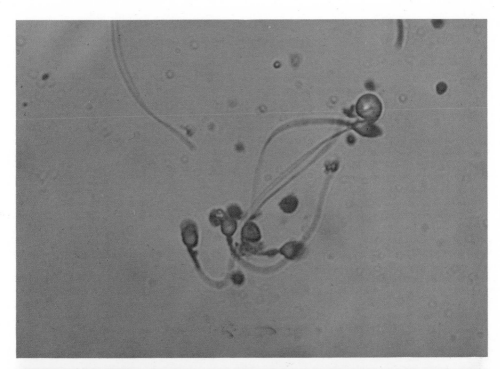

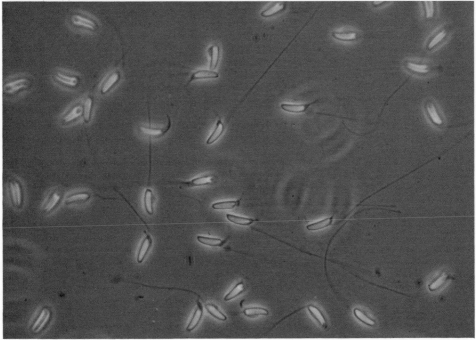

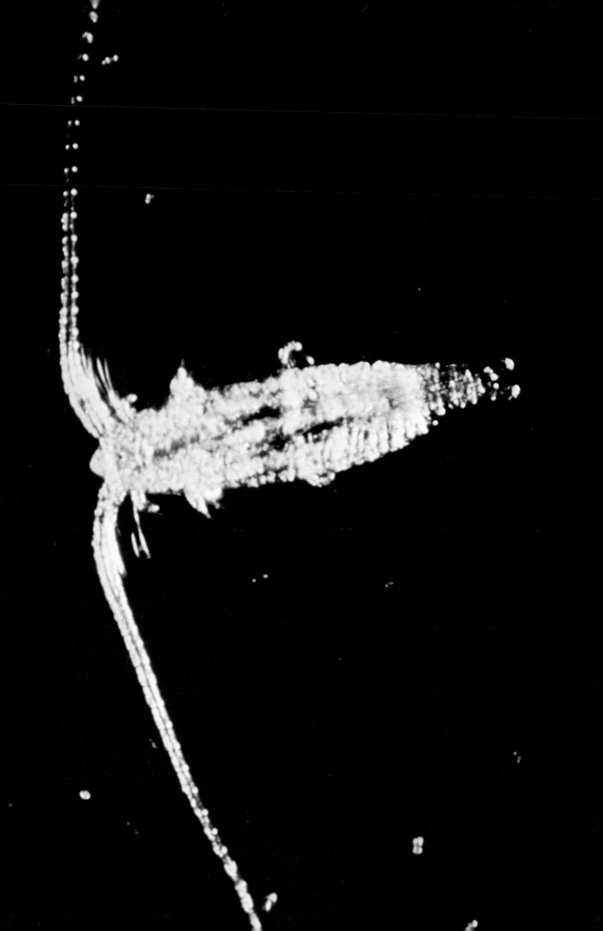

(above) Fractured femur of a cow. Magnification x42, zooming in to detail *(right)* at x1,240.

(facing)
The marine copepod Calanus, the 'insect of the sea'. These minute shrimp-like organisms weigh as little as 1/140,000th of an ounce. In the deep ocean they can grow to an inch in length. They manufacture their own supply of wax which is used for food during periods of starvation. The copepod is the first animal in the marine food chain. It lives on microscopic plants – phytoplankton – and converts their fat into polyunsaturated liquid wax which is stored in an oil sac. It has been estimated that the copepod population off the coast of California alone contains some 800,000 tons of wax.

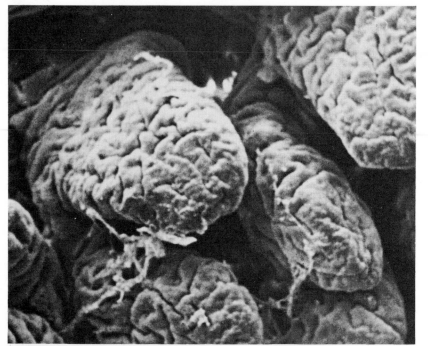

The villi of the normal human intestine. They are responsible for absorption of food as it passes through the gut. They measure about a tenth of a millimetre across and are covered by closely packed short stubby rods (too small to see in this picture) that increase the surface area for maximum absorption. (mag. x264.)

Pass by a smoking factory chimney and you will probably breathe in particles like these. They are tiny grains of silicates, each less than 1mm across. They are formed when pulverised coal is blown into a furnace and some of the molten mineral fragments in the coal go up with the smoke. (mag. x90.)

(facing) Calcium carbonate — the 'fur' inside your kitchen kettle. (mag. x2,740.)

(below) Niobium crystals. Ten years ago titanium was the new wonder metal. Today it is niobium. Its full potential has yet to be explored: it has a high melting point of 2,400°C, is electrically a super-conductor, and is highly resistant to chemical attack. So far it has been found in Canada, South America, and Russia. (mag. x2,900.)

(facing) Seen under any kind of optical microscope at this magnification (mag. x25) a crystal of silicon would appear to have a smooth, highly polished surface. But using a highly complex technique in a scanning electron microscope developed by Douglas Coates at the Royal Radar Establishment, Malvern, a pattern of lines becomes visible which gives an instant clue to the quality of the crystal. As silicon is the material used in micro-miniaturised electronic circuitry in computers and space satellites, it is vital information. The diamond pattern in the centre of the picture is silicon's 'thumbprint'. Every crystal of silicon has this mark, making it instantly recognisable. Other materials have their own prints. But they show up only when Douglas Coates' new technique is used.

(bottom right) Crystals of common salt as seen with a scanning electron microscope. Note the great depth of focus. (mag. x800.)

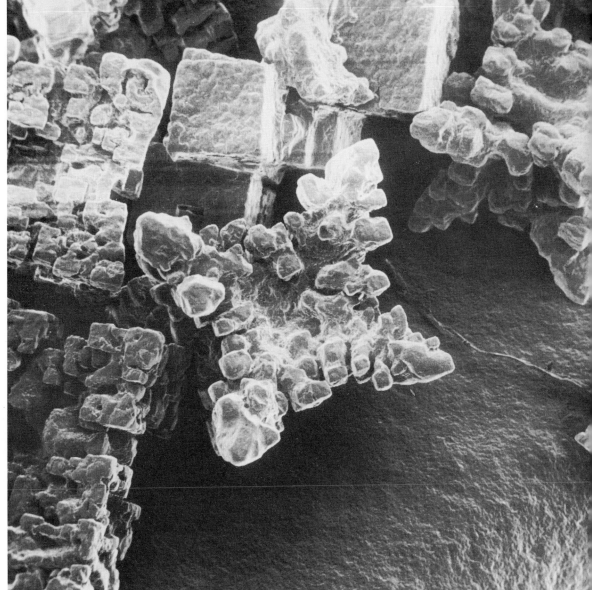

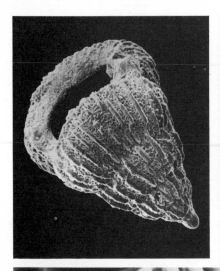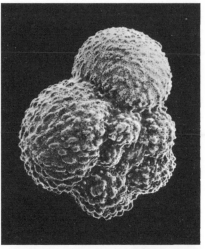

Fossils of tiny marine animals between 80 and 100 million years old. They were found embedded in rocks in Jordan. Fossils like these are used by geologists to indicate whether there are likely to be oil deposits in the search area. By examining samples of rock as the drilling progresses, geologists can tell from the fossils at what point in history the rock was laid down and from that deduce if there is oil to be found. (*left* mag. x136, *right* mag. x154.)

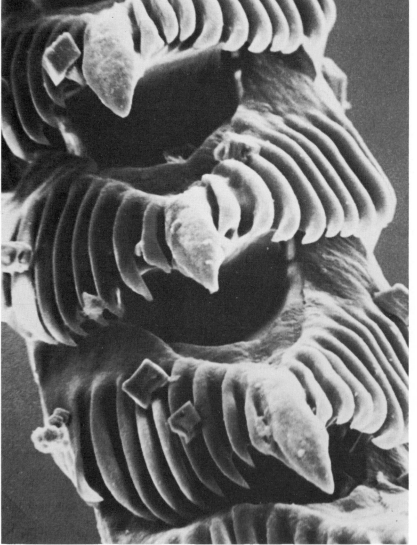

Remarkable tongues. Both come from tropical sea slugs and because their diets are different their tongues are different too. *(facing)* The feathery surface of Rostanga Arbutus helps the slug to cope with its staple food — sea sponges. (mag. x3,600.) *(left)* Because Pteraeolidia Semperi feeds off the sea anenome its tongue is covered with horny scales that puncture its victim to extract the juices. (mag. x3,100.)

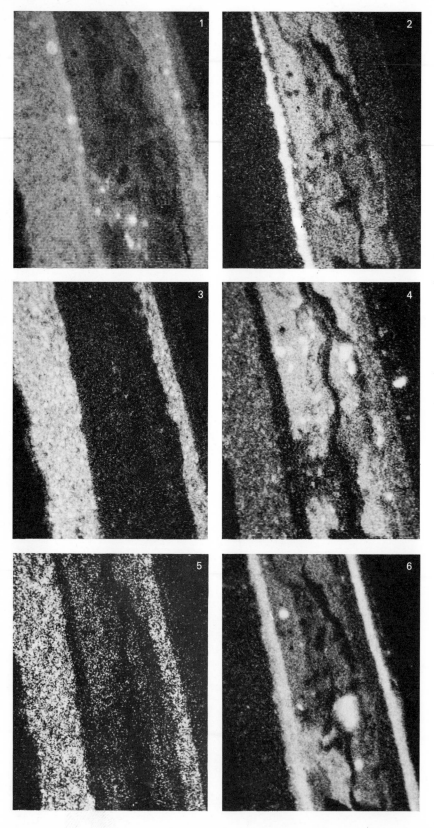

These six pictures show how the scanning electron microscope can be used for analysis. In each case the sample is ship's paint and the pictures show the patterns of emitted radiation of various elements. What happens is that when the sample is bombarded with electrons, X-rays are given off. These are analysed by a built-in spectrometer which measures the different wavelengths — each element having its own wavelength. The technique is called electron probe microanalysis.
1 Chlorine
2 Copper
3 Lead
4 Silicon
5 Magnesium
6 Iron.
(mag. x53.)

The common midge.
These minute flies
swarm in northern
areas of the temperate
regions. They are
particularly abundant
in parts of Scotland.
In this picture the
midge's left antenna
has been removed to
give an uninterrupted
view of its compound
eye. (mag. x60.)

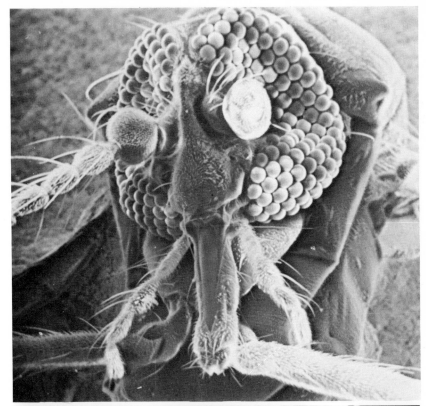

A close-up of the
tip of the mouth parts.
Within the sheath of
the top lies the
styleto that pierces the
skin so that the midge
can suck blood. (mag.
x1,250.)

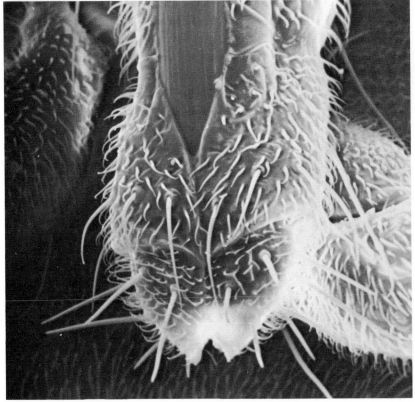

J

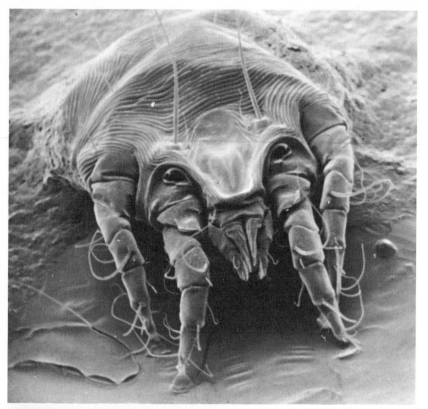

The Mange Mite (Otodectes cynotis). Here is the female of the species which causes mange on cats, rabbits, and rats. In cats it is usually found on the head, especially inside the ears. Fortunately it is not able to establish itself on humans. (mag. x260.)

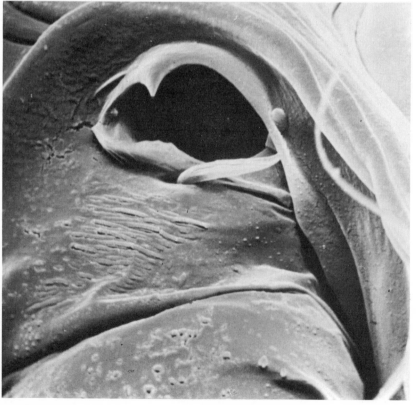

The mite is blind. Where eyes occur on many animals there are openings to glands. This is a close shot of the opening to the supra-oxal gland *(bottom left)*. (mag. x1,560.)

(facing) Caloglyphus — a tiny mite that lives in damp mouldy grain and other foodstuffs. These pictures were taken on a scanning electron microscope. 1. General view of the mite's mouthparts. (mag. x1,930.) 2. Closer view (mag. x4,030.) 3. Close-up of the cheliceral teeth that do the damage. (mag. x9,650.)

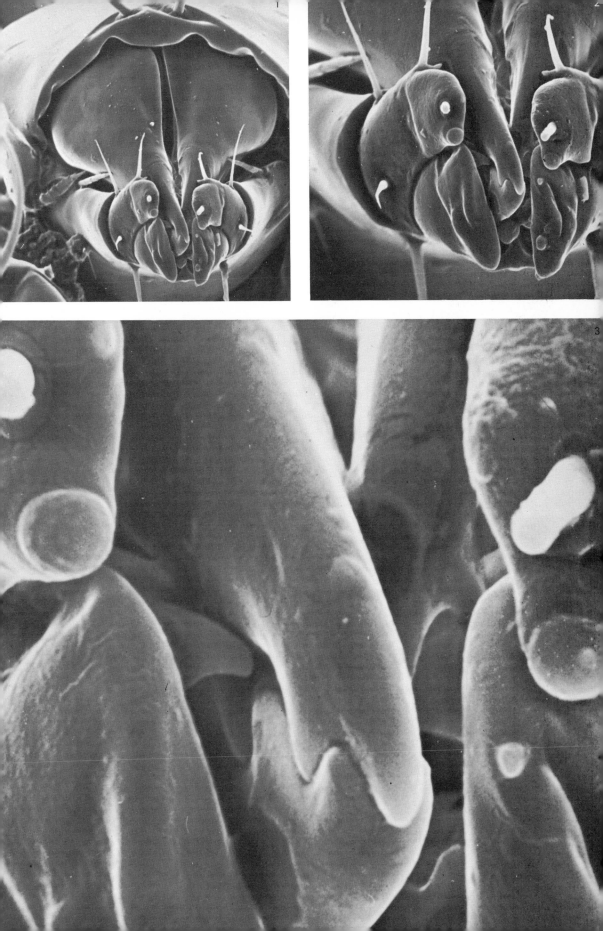

(far left) Ctenoglyphus plumiger. Though it looks like a beautiful sea creature, it lives in the general debris around the base of haystacks. The purpose of the elaborate feathery setae is not known. It may distract other predatory mites from making a successful attack. (mag. x243.)

(bottom left) Close-up of its mouthparts. (mag. x3,110.)

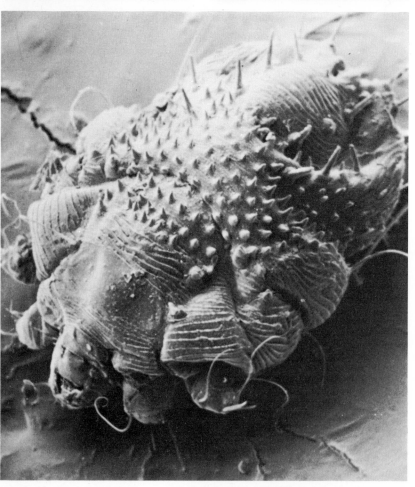

The Itch Mite (Sarcoptes scabiei). This is the creature responsible for scabies in humans, a disease which is at present on the increase. The mite burrows through the horny outer layers of human skin and is admirably adapted to its habitat. Its legs and mouth-parts are well protected and the spines sticking out of its back probably assist its movement through the burrow. (mag. x322.)

A cow parsley seed —
the plant that grows
up to 2 ft high in
British hedgerows.
The reason for the
spikes is not fully
understood as the
seed is distributed by
the wind and not by
animals and birds.
This is a good
example of how the
scanning electron
microscope can
'zoom' in to the detail
of a specimen.
(1) Magnification
starts at x40
(2) increasing to
x430 and (3) to
x2,040.

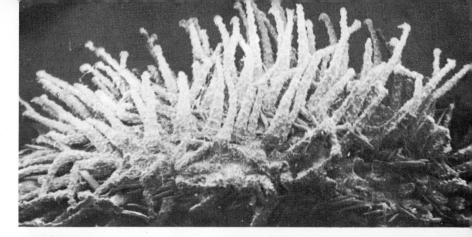

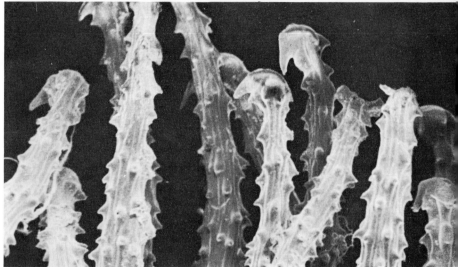

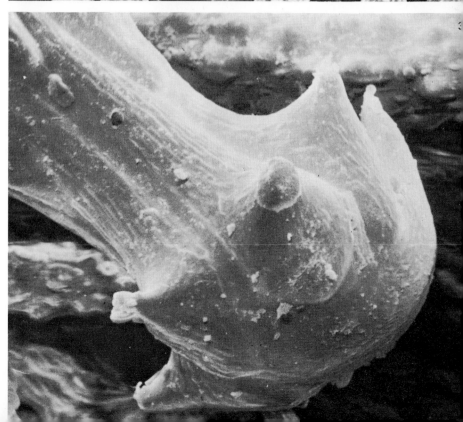

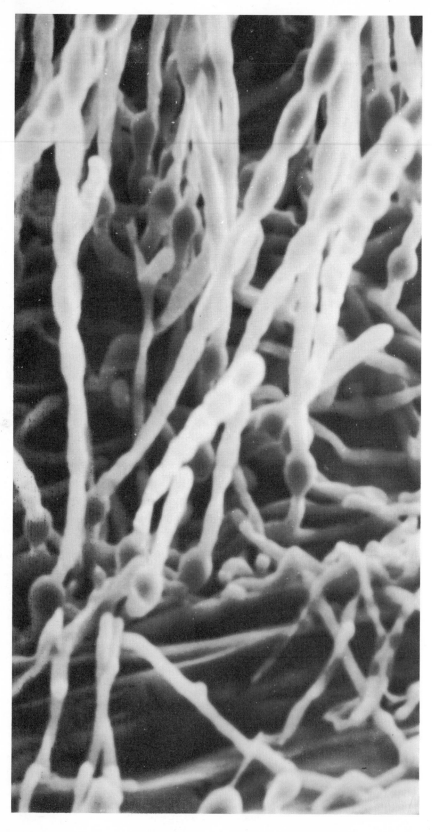

(left) Powdery mildew spores on Kentucky blue-grass. The mildew, erysiphe graminis, is a parasitic fungus that does tremendous damage to wheat and other grain crops. Attempts are under way to breed varieties of corn that are resistant to the damaging spores. So far, scientists have not had much success. (mag. x680.)

(right) The underside of a geranium leaf. The hairs with globules at their tips are oil-producing glands. The slits in the surface of the leaf are stomata — the holes through which the leaf breathes. (mag. x250.)

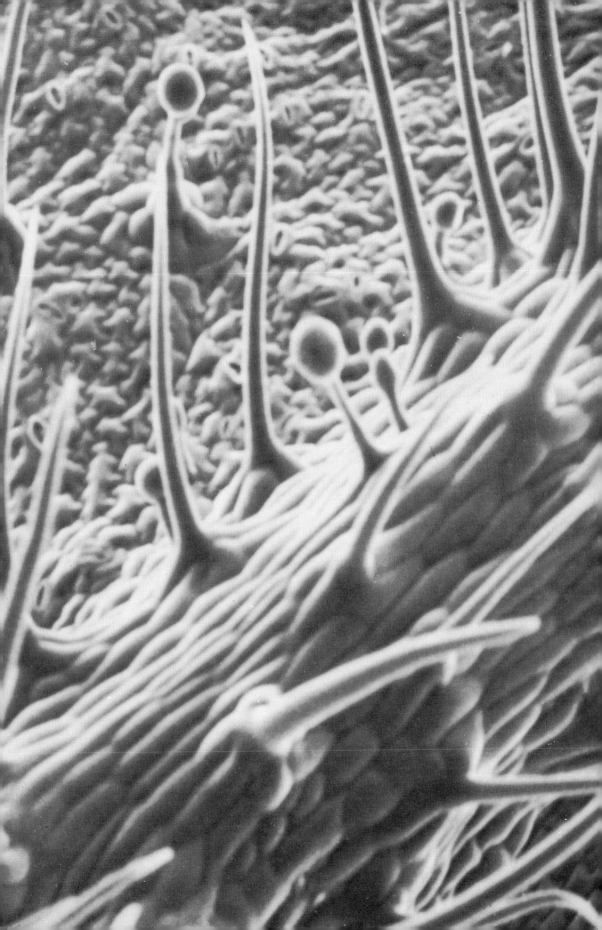

Can you find your way around the Inner World?   Is it a . . .

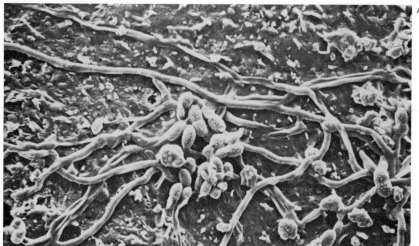

A 1 Live sperm of the
giant land crab
2 Fungus and spores
on the surface of a
sturmer apple
3 Germinating
mustard seed
4 Interior of a cancer
cell

B 1 Crystals of lead tin
telluride
2 Magnified
computer circuitry
3 Common salt
4 Pollen grains of the
passion fruit flower

C 1 Tip of a human
    tongue
  2 Egg cluster of the
  common stickleback.
  3 Stomach lining of a
  Thompson gazelle
  4 Moth's wing

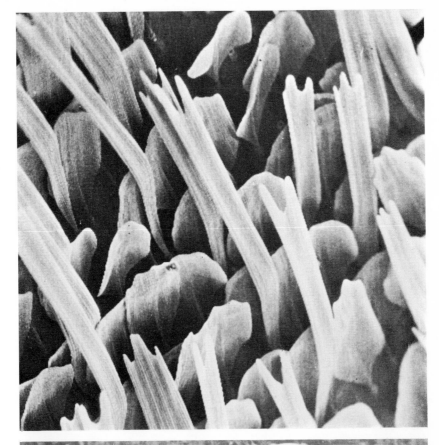

D 1 Glass particles
    found in Moon dust
  2 Fat globules in
  lipstick
  3 Powder grains of
  tin bronze alloy
  4 Influenza virus

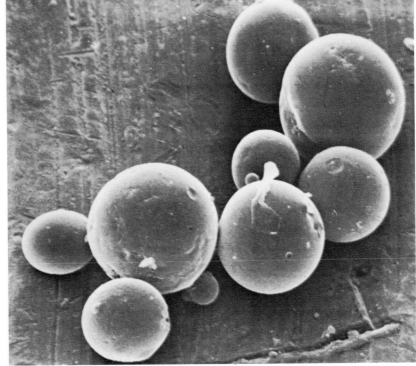

Answers
on page
157

F 1 Deep sea snail
  2 Cheese mite
  3 Giant sloth
  4 Wasp's head

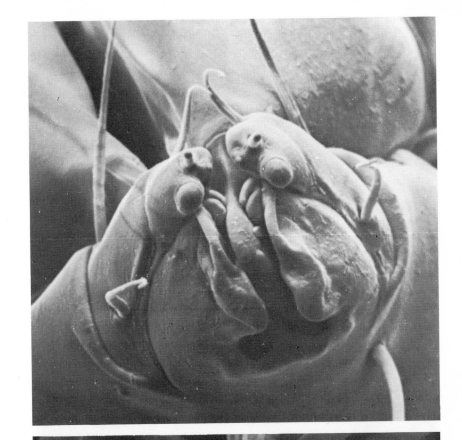

G 1 Epsom salts
  2 'Fur' from a kettle
  3 Ruby crystals
  4 Teeth of the
    feather mite

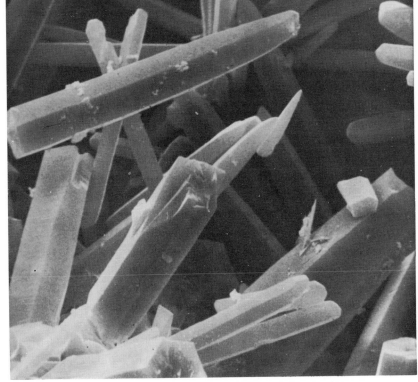

(facing)
E 1 Tip of a hypodermic
   needle
  2 Section of carotid
   artery
  3 Bee sting
  4 Transistor
   connection

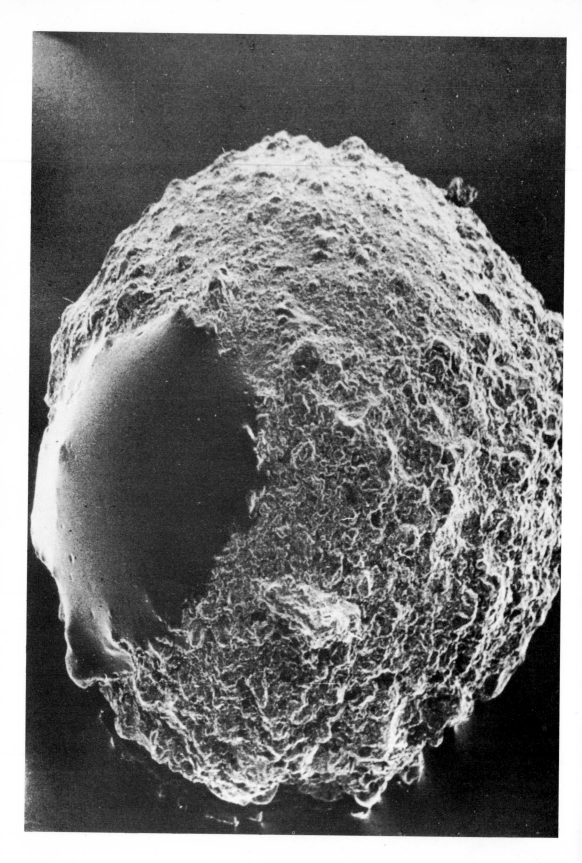

H 1 Stress marks on a
human toe nail
2 Nerves within a
rat's brain
3 Cutting edge of a
diamond drill
4 Silicon crystal

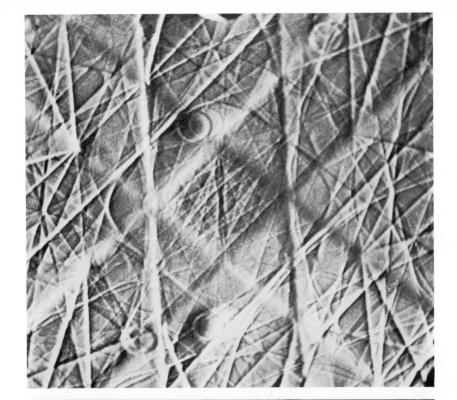

I 1 Grains of icing
sugar
2 White blood cells
3 Polyurethane
particles
4 Sodium chloride
crystals

(facing)
J 1 Sea urchin egg
2 Human liver cell
3 Glassy droplet
coated with Moon
dust
4 Portland cement

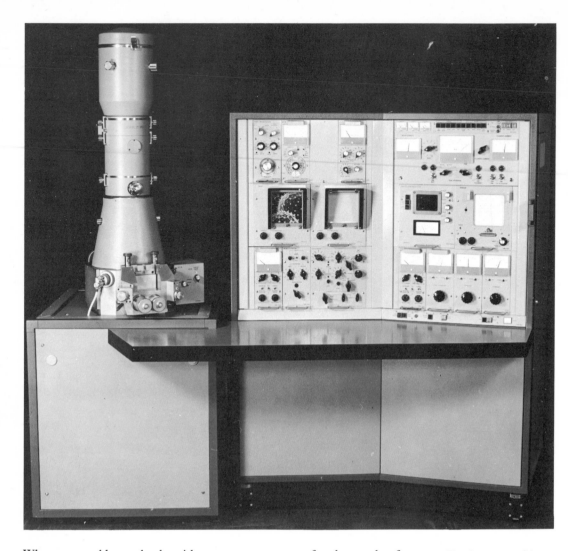

What you could see clearly with your own eyes was for thousands of years accepted as an accurate indication of precisely what was there. Then, towards the end of the 16th century the first compound microscopes were invented and it became clear that in scientific terms our unaided eyes were capable of showing only a coarse version of reality. Four hundred years went by before the electron microscope came on the scene. In its turn it opened up a new world of micro-shapes and structures. And a few years later the *scanning* electron microscope was developed.

Like the conventional electron microscope, it works on the principle that if you fire a stream of electrons through a vacuum at an object it is possible to capture them as they bounce off, beam them onto a television screen, and reveal a magnified image up to 100,000 times bigger than the original. What makes the scanning electron microscope rather special is its ability to produce images with an unusual depth of focus – some 300 times greater, in fact, than can be achieved by the best light microscopes. This makes it an ideal tool for examining rough surfaces.

Much of the art in this kind of microscopy lies in knowing how to prepare the samples to be viewed. Though living insects have been examined for short periods within the vacuum of the microscope, biological samples usually have to

The Stereoscan S4, one of the most sophisticated scanning electron microscopes in the world. The column containing the electron gun and the sample to be viewed is on the left. Electronics and viewing screens are on the right. With a modern microscope like this you do not have to be an expert to get good pictures. Most of the operation is automatic. But you do have to be an expert to interpret correctly what you see.

be specially treated to make them 'visible' to the electron beam.

For example, in order to see red blood cells, they have to be smeared on a glass slide, carefully dried, then covered with a very thin film of gold and palladium which is evaporated on to a thickness of a few molecules only. This enables the cells to conduct the electron beam more efficiently. The result is an incredibly clear and detailed picture with an almost three-dimensional quality. Nowadays it is even possible to see deep inside a biological cell by first carefully etching away its surface with an ion-beam.

Of course, with metals and minerals that conduct electricity, there is no need to evaporate special coatings onto their surfaces. The samples are simply placed inside the microscope and viewed as they are. At the same time experiments may be carried out and you can actually watch what is happening at the moment it occurs.

But perhaps the most exciting feature of the scanning electron microscope is its ability to 'zoom' in to a specimen. By gradually increasing the magnification while carefully manipulating the object it becomes possible to literally 'crawl' into the innermost recesses of a human cell or a particle of Moon dust and burrow away like a potholer in an unexplored subterranean world.

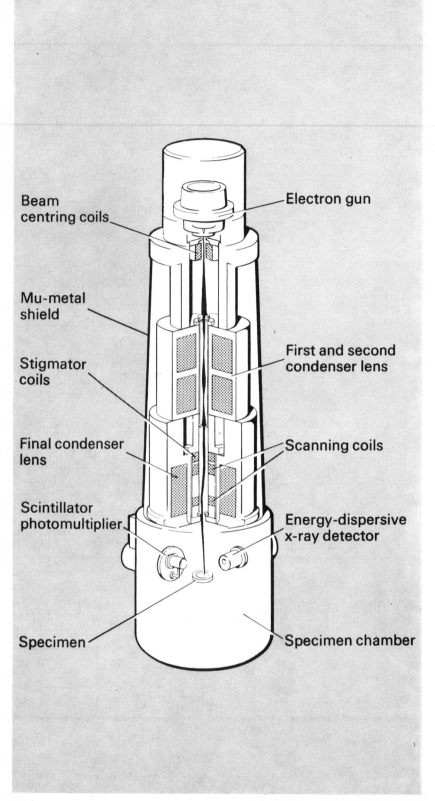

Beam centring coils

Electron gun

Mu-metal shield

First and second condenser lens

Stigmator coils

Final condenser lens

Scanning coils

Scintillator photomultiplier

Energy-dispersive x-ray detector

Specimen

Specimen chamber

The diagram shows the layout of the electron optics of a scanning electron microscope. Electrons emitted by the electron gun are focused onto the specimen which is fixed to a mounting pallet in the chamber at the base of the column. The beam scans the specimen to produce a magnified image which is presented on a television screen not shown in this drawing. Magnification can be changed at the flick of a switch and recording is simply a matter of photographing the TV screen.

Answers

A2
B1
C4
D3
E1
F2
G2
H4
I4
J3

Scoring: One point for each correct answer

10 pts: You are in the professional class in Inner World travellers.

9 – 7 pts: Without doubt, you feel at home in the Inner World.

6 – 4 pts: You need a guide to some parts of the Inner World.

3 – 1 pts: Take care. You get lost too easily in the Inner World.

No score: Your passport to the Inner World is invalid.

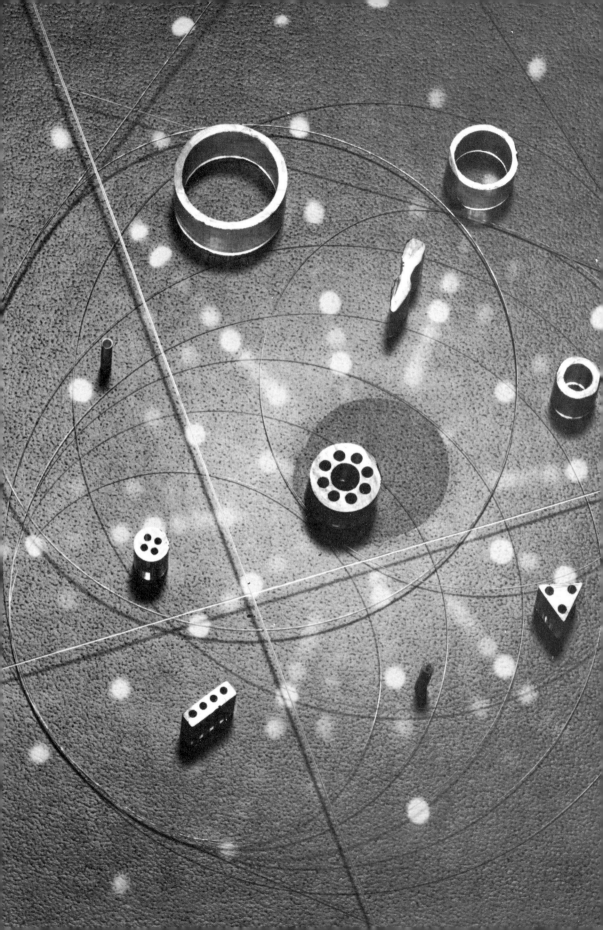

7　Tools of Change

It is fashionable today to look ahead with a sense of despair to a world without liberty, privacy, silence or green grass. Leading journals carry poems written by children who wonder if there will be birdsong for *their* children to hear. Pollution hangs over the cities in movies, books and politicians' speeches. And in every one of them the cry is to stop the technologists before they finish us all. Unfortunately, the very technology used to put that cry on the drugstore shelf, the processing technique for developing the film, the microwave link that carries the politician's battle-cry home to the voters – are all an indication that it is too late to talk in terms of 'stopping' anything. The need is not to stop, but to change direction.

There has never been a scientific or technological development that did not bring with it its own means of control. There is no need to look for new tools and materials to bring about the change – we have the tools and materials already. Technology produces them faster than it makes anything else, because it is with them that technology changes our lives. The tools and materials are, if you like, the only neutral products of scientific development. They are as constructive or as dangerous as the people who use them.

But it can often take years for these tools and materials to get down to a level where they make any difference to the man in the street. Mostly their use is limited, in the initial stages, to military or industrial use. For reasons of state or commercial security, nobody gets the chance to develop them for the enhancement of everyday life. Advanced radar systems that would be invaluable for civil airlines are used first to defend our shores from attack, until such times as the system is superseded by a better one. Until then the civilian airline passenger perhaps travels in less safety than he need. We are only now beginning to enjoy the benefits of the 10-year-old research and development that went into putting a man on the Moon. While we in the West struggle with the problem of what to do with the technological output, we are busy exporting some of it to underdeveloped countries. If it is difficult for *us* to comprehend the meaning of the changes it brings, what about *them*?

Tools and materials are the means by which we can reach out and change the world around us in the most basic of ways. In most cases they help to carry out four fundamental activities: explore, construct, test, and decide where to do all three.

One of the latest devices for the explorer was developed by Westinghouse. It was first tested out in Vietnam. With it you search out your enemy, and follow him without once having to refer to a map along the way. When you need to know where you are the device tells you. It's called the Improved Position Locator and it's the first time electronic navigation systems have been successfully adapted for the individual on foot. The system consists of a pair of boot antennae, a backpack computer, and a display and control unit. The antennae are used to measure the length of each stride. In the backpack there is an electronic compass that tells the computer in what direction the steps are going. The direction is multiplied by the length of step, and the results are stored in a counting device. Before the mission begins, the known co-ordinates of the starting point are fed into the computer, and all measurements it makes thereafter refer back to them. To find out where you are switch on the display unit and a pair of counters gives you your position in 'eastings' and 'northings' which can be related to the map of the area.

Saphicon – nearly as tough as diamonds. It can be 'grown' to any shape. (See p. 167.)

The essential information on your length of stride is determined by the boot antenna in the following way: one antenna is a receiver and the other a transmitter. As the feet separate on full stride, the signal strength coming into the receiver weakens. This strength level is fed into a circuit in the backpack that is programmed to know at what distance certain signal strengths will occur. Hence it knows how far apart your feet are, and therefore what distance you are moving with each step. This distance is related to the direction you are heading in by referring it to the compass heading at the time one foot passes the other. It is at this point that the body is generally facing the direction of travel. The problem of going over hills, which could be taken by the computer as steps along level ground, is solved by dividing hilly terrain into three categories according to steepness, and superimposing a correction factor into the calibration dial used for setting the display counters. It has been shown that this method keeps the error factor within 2%. The benefits the Position Locator will bring to surveyors and planners, particularly in developing countries, are tremendous.

Once you have arrived at your destination, another new device will help with analysis of what lies under the surface. The detector operates on low-frequency radio waves which penetrate the upper layers of the soil and bounce off any underground obstructions they meet. Previous detectors would only indicate metal, this one can be calibrated to indicate any material. It consists of a low-frequency radio transmitter, a receiver to pick up the returning 'echoes', and a tape-recorder on which the echoes are recorded as a series of blips. The operator criss-crosses the area, switching the tape off as he reaches the end of one traverse, and on again as he begins the next. When the area has been completely covered in a series of parallel traverses, the tape is connected to a print-out system. As the trace unfolds, all the underground obstacles show up as ink blobs, whose size is related to the length of the original echo on the tape. So far the detector works only up to 4 ft underground, but already a more powerful version is planned. With this device the ease with which ground can be prepared for building should considerably reduce both time and cost of construction. Whether or not it will reduce the price you pay for the finished product remains to be seen.

The ability of low-frequency radio waves to travel through the earth has been used recently in the design of a system for locating men trapped underground. At the surface a low-frequency aerial several thousand feet long is laid out along the ground. It permits rescuers to warn men first of danger or of an accident that has happened elsewhere in the mine, for each miner has a tiny radio receiver in his lamp battery case. But in order to make sure that the miners stay alive after the accident the survival system also includes two shelters. A small auxiliary shelter moves with the men at the face as work advances. Six steel sections, or modules, each with its own set of wheels for moving from place to place, and two end-bulkheads go together for the entire structure. Pads mounted on the bottom of the bulkheads bolt to the floor to keep the shelter from being lifted and hurled down the tunnel by the explosion. Oxygen is provided by burning chlorate candles, which produce no open flame and very little residue, while giving off high quantities of the gas. Inside the auxiliary shelter 15 men can live for 2 weeks.

These candles are also the mainstay in the miner's personal life-support system. Up to now breathing systems filtered the carbon monoxide out of the air the miner was breathing. The new system removes him from all contact with the local air. A chlorate candle fills the mylar plastic hood he wears over his head with enough oxygen for him to carry out heavy work-loads. Since his oxygen needs will vary, he is always provided with the maximum amount he could use and, to cope with possible excess pressure, a venting valve is set into the hood. A rubber neck seal makes sure that no air will escape, and none will enter from the contaminated atmosphere outside. Most important, chemical filters of the type used in the

A computer to find yourself. The pack on his back will automatically keep track of every step no matter where he roams — and provide him with information on his location at any instant.

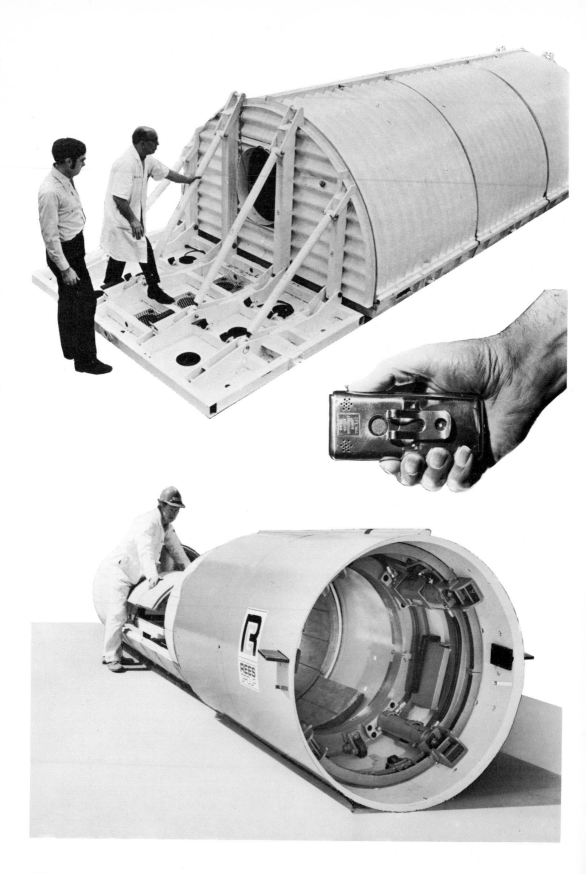

Apollo spacecraft, operating on lithium hydroxide, are used to remove carbon dioxide from the air in the hood, so that the small amount of oxygen exhaled with the carbon dioxide can be used again.

The other, main survival shelter consists of a 75-ft tunnel cut into the rock with blast walls at each end. An 8-inch air pipe connects the chamber with the surface, for dropping supplies, carrying air, communication links and electrical power. The men trapped inside this shelter will be able to hear voice messages from the surface but will only be able to answer with a series of coded pulses sent out by a small transmitter. These coded signals will activate lights at the surface indicating that the push button responses 'yes', 'no', 'unknown', 'repeat', 'good' or 'bad' have been keyed by the trapped man. The reason they will not be able to talk to the men at the surface is that the battery power-consumption involved would be prohibitive.

To find anyone trapped alone the system uses tiny ground microphones, called geophones, sensitive enough to pick up the sound of a shovel against the wall of the mine from a distance of 2500 ft. During rescue operating several geophones receiving signals of movement underground route their information through the local terminal of a computer containing a detailed map of the mine. In tests so far the sound sources have been pinpointed to within 12 ft. As long as men have to go underground, systems like this should make their life safer, if not more congenial. That will have to wait until technology does away with the need for manned mining at all.

For those mines having to wait to install the Westinghouse system – and small uneconomic mines could wait for ever – a new development by the British Mines Research Establishment could give the individual miner at least warning that the air around him was low in oxygen. It all depends on the behaviour of an electrical current. The basic warning system consists of an electrolytic cell – that's a small container of salt water with a silver wire cathode in the water. A hole in the cell's wall is covered with a metallised plastic membrane which is in turn connected to the anode fixed on the cell wall. Now the amount of electricity flowing between the anode and the cathode, each of which is connected to the opposing poles of a battery, depends on the amount of oxygen getting into the salt water. And the oxygen can only get in through the porous metallised membrane. As the oxygen level drops, the current running through the water is disturbed and this in turn activates warning lights or buzzers. The whole system has been miniaturised to go into small, clip-on warning units, that can be used not only down mines, but anywhere the pressure of oxygen is vital: a hospital incubator, an intensive care unit, an oxygen tent. One day your life could be saved by this tiny bleeping noise first developed to save a man from death hundreds of feet below the ground.

More often than not it is the man working only a few feet from the surface that concerns us the most. Every year thousands of hours' delay is caused in cities as pipes are laid. Now a British company has come up with an idea that may take the pipelayers out of sight and hearing for good. They've developed a portable mini-tunnel that could have application in every area of underground work from pipelaying to putting down foundations. Up to now similar kinds of semi-automated tunnellers had been available down to 54 inches in diameter, and the reason they haven't already made life quieter in your neighbourhood is that most of the pipe systems in a city are smaller. This new mini-tunnel will permit the laying of pipes almost any diameter below 54 inches.

The system consists of a shield, inside which the excavating work takes place. It is hydraulically driven forward as the earth and rock is taken out. Every 2 ft it advances tunnel-lining sections are brought up and mounted. They are made of concrete in the shape of a segment of a cylinder wall around an arc of 120°, so that three of them fitted together made up a 2-ft section of the new tunnel wall.

A miner, working within the shield, erects one of the three 120° concrete segments which form a complete tunnel ring 2 ft long.

The shield then advances gradually until the next section can be mounted. While this is happening gravel, grouting or other suitable material is being injected around the outer skin of the concrete sections to pack the new tunnel firmly in place.

Of course many small-diameter pipes must still be laid in trenches, although increasingly now engineers are able to lay large-diameter pipes and route several smaller pipes through them. The difficulty with these small pipes, whether buried individually or collectively, is that examining them has been difficult. Now a British company has developed a way of literally going down the pipe and looking at it from the inside, throughout its length. The crawling eye that does the work is the world's smallest commercially available television camera – called the Falcon. Using a black and white videcon tube producing high-definition pictures, the entire camera and its casing is only $1\frac{1}{2}$ inches in diameter. The image-producing tube itself is reckoned to be the smallest in the world, only $\frac{1}{2}$ inch across. The lens gives a field of view covering an arc of about 108°. The body of the camera is $5\frac{1}{2}$ inches in length, and will easily negotiate a $2\frac{1}{2}$-inch pipe bend with an average radius of turn of 12 inches. Negotiation of these tiny pipes is done by mounting the camera on a small, wheeled rig which can be hydraulically expanded until the rig wheels touch the sides of the pipe. Once in this position, with the camera in the centre of the pipe, the whole unit is pulled through by cable. If close-up inspection of the sides of the pipe is needed, a rotating mirror can be mounted ahead of the lens and angled to show the pipe walls at 90° of the lens's field of view. The scene you see on the television screen of the surface is lit by tiny bulbs set in the case around the lens itself. Once the engineers spot a section of pipe that needs close examination the pictures can be recorded on videotape. Industry has been quick to see the possibilities Falcon presents, and already the camera has been used to examine offshore cables in Spain, in North Sea offshore oil-drilling rigs, and in food-processing plants. Anywhere small tubes need examination, the camera will go.

It was discovered a few years ago that if you took concrete and introduced fibre into it, you produced a composite material that had all the strength of the fibre together with the rigidity of the original concrete. It had strength and, above all, lightness. Probably the best known of these composites are asbestos cement (cement reinforced with naturally occurring fibres) and fibreglass (polyester or epoxy resin reinforced with glass fibres). From a constructor's point of view the asbestos cement is inexpensive but weak; and the fibreglass is strong but costly. So recently, at the British Building Research Station, experiments have been conducted to try to put fibreglass and cement together. The trouble was that cement contains alkali, which attacks most types of glass, preventing the fibres from integrating well with the cement. But at the Corning Glass Works in America a type of alkali-resistant glass has been produced for components in steam lines. A special technique was developed for turning the glass into fibre, which formed a 'mat' on which the cement was laid, producing a mix of about 95% cement and 5% glass. The result has been a new composite that is extremely hard yet flexible. It's weatherproof, and can be sprayed over simple moulds to form cladding units, or panels for the permanent shuttering of concrete beams. Most important the new material is relatively cheap to make and extremely easy to handle.

It is the handling of many chemical products that has kept them out of the engineers' hands so far. In most cases the degree of control and skill needed during their application is simply not available among the unskilled labour on a construction site. Wider use has had to await the development of simpler handling techniques and materials. One of the most versatile materials to suffer from these constraints has been fibreglass-reinforced plastics, which have been in limited

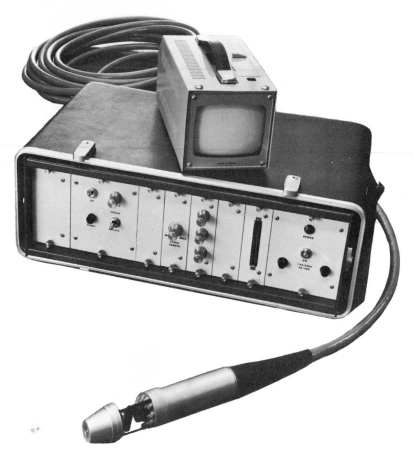

The miniature Falcon television system. The camera *(foreground)* has a diameter of only 1½ inches. It has a built-in lighting system and can be focused and rotated by remote control. Its purpose: to travel down small-bore pipes to examine them. (See page 165.)

use for some time in the car industry, in chemical storage-tank production, in fishing-rod manufacture, and so on. But the plastic used to reinforce the fibre-glass had always to be weighed and mixed very carefully under special conditions. Now an American company has simplified the process. The material looks like a mat; it is made of chopped fibreglass sections on to which a polyester resin is poured. The resin hardens to provide as flexible a substance as is required. The ease of handling involved lies in the composition of the resin itself. Before it leaves the factory a number of hardening agents are added which form a stable coating over each individual glass fibre. It is this pre-treatment of the resin that allows anybody to use it. It can be sprayed, rolled or brushed on to the mat. In a matter of minutes, in whatever weather conditions, the mat hardens and can be used for laying temporary roads over difficult terrain, landing pads for VTOL aircraft or helicopters, floors for silos, as well as for temporary repair work to swimming pools, storage vessels, or even the roof of your house.

Another new material, of particular value to the aircraft industry, is Boron fibre. Boron is an element that melts at about 2300°C. The fibre is formed by allowing a mixture of Boron Tichloride and hydrogen to come into contact with a heated tungsten layer. The fibre is then vapour deposited on to the tungsten. At the moment it is extremely expensive – about £100 per pound, but according to the Avco Corporation who developed it, the price should drop by the late 1970s to about £25 per pound.

The main use of Boron fibre is for reinforcing aluminium composites. These are less strong longitudinally than Boron fibre, but their transverse and sheer stiffness is higher and they can maintain their quality at high temperatures. For this reason the fibre is aligned longitudinally on such structures as beams, stiffeners, tubes, rods, etc. Avco claim that in Boron fibre they have a material of greater tensile and compressive strength than carbon fibre – which caused such excitement a few years ago – and it shows weight-savings of some 25% for equivalent strength or stiffness. It is early days yet to say whether carbon or Boron will prove the more effective, but it is fibres like these that hasten the day when you may take off on your holiday in a 1000-seat aircraft that will make the Jumbo jet look small.

Cost is always the dominating factor in any industrial development. No matter how good the material is, how strong and versatile, if you can't make a profit because it costs too much to use it, then it has a limited future. This used to be particularly true of some of the rarer materials like diamonds and sapphires. But now it looks as if sapphire, a material nearly as tough as diamonds, may suddenly have tremendous industrial possibilities. A new sapphire material called Saphikon has been developed that can be grown to almost any shape or length. This unique growing process increases the already high strength and durability of the sapphire, because the grown material is free from machined surface irregularities and grinding micro-cracks which can lead to failure. Its chemical inertness, extreme heat-resistance and immunity to abrasive corrosion has led to new tube designs, from delicate medical instruments to sand-blasting nozzles, for a sapphire tube will take virtually any corrosive liquid. The finished product consists of one giant crystal that can be grown in any direction; this single crystal-structure gives Saphikon zero porosity, which prevents leaking and vaporisation. It also has extreme transparency over a wide wavelength from deep ultra-violet to well into the infra-red frequencies.

The transparency of glass has always been of cardinal importance in spacecraft. All solar-powered satellites use arrays of light-sensitive solar cells which react to sunlight and so produce electricity. In the space stations due to be launched in future, conditions on board will depend entirely on the availability of power supplies. Any degradation in the ability of the solar cells to produce electricity is dangerous. It is for this reason that the solar cells powering satellites at this moment swinging around the Earth in orbit have protective covers about the size of postage stamps and only a few thousandths of an inch thick. They are made of glass carrying a multi-layer coating to protect the cells from radiation and micro-meteorites. There are sometimes as many as seventeen of these expensive coatings. The problem with them is that because they are laid on one layer at a time, they eventually delaminate. But recently the Pilkington Glass company in Britain has developed a new kind of glass that is not only extremely transparent but is also an effective radiator of waste heat. During development various additives were made to the basic mixture of sand, limestone and dolomite from which glass is made. Then it was found that the addition of a small amount of cerium oxide protected the glass from discoloration by radiation, so eliminating the need for a multi-layer coating. One coating, common to all solar covers, is, however, still added by vacuum deposit. It is a layer of magnesium fluoride which acts as an anti-reflection agent, and therefore as a further aid to the transmission of the sun's light through the cover.

At present the covers, which are lighter, cheaper, more reliable and have a longer life than conventional types, are being produced for the ESRO European satellites. If they prove to be as effective as the developers claim, they could contribute significantly to the orbital life of a space station and to the safety and efficient working capacity of the men and machines on board.

There can be few transparent materials that are more immediately vital to many people than the plastic from which a contact lens is cut. The number of contact-lens wearers grows every year, but for some people a hard plastic button in their eye is too uncomfortable to bear. This problem looks like being eliminated with

New glass for space. Unlike other glasses used in satellites, this one does not discolour when bombarded by solar radiation. (See page 167.)

(below)
A contact lens as soft as silk. It is made of a new material that actually absorbs tears to stay pliable.

the recent development of a soft lens. It is made of a new kind of material – hydrophilic plastic – which absorbs liquid, becomes soft, and is extremely easy to both fit and wear. The lens is cut from a blank on a lathe but is slightly larger than a conventional contact lens and fits under the eyelids. As the lens lies on the eye, the natural reaction produces tears which are absorbed by the plastic, allowing it to remain permanently wet and soft. One other advantage of the new plastic is that because of its extreme porosity, it permits air to pass through to the eye at all times.

One exciting possibility this new material presents is its potential use as a new kind of eye bandage. Used to cover diseased corneas it could protect the area from outside interference, and at the same time do no damage to the affected tissue. Experiments have also been conducted using the lens as a slow-acting source of medication. If the plastic is impregnated with certain drugs, it will release them slowly to treat an eye condition over a considerable period of time. Medical researchers are extremely excited by the potential uses of the hydrophilic lens,

Many blind people might see again if it were possible to replace the damaged lenses in their eyes. Now an operation devised at Southend Hospital in Essex makes this a reality. It is a three-part procedure. The first operation is designed to strengthen the damaged cornea. Then a few days later, a 4-millimetre hole is cut into the cloudy lens, the damaged tissue is removed, and a tiny plastic lens holder is carefully inserted.

A few weeks later when the eye has healed the third stage can begin. During a 15-minute operation the surgeon screws a plastic lens complete with artificial pupil into the holder already embedded in the eye. The operation is so delicate he does not wear gloves for fear that they might impair his sense of touch. A day later the patient can see again.

1 Start of operation. The lense holder *(left)* and plastic lense *(right)* that will restore sight. Note the thread that ensures the lense remains firmly in place *(left)*.

2 The holder is carefully pushed into the eye.

3 Finished operation. Plastic lens with artificial pupil safely in place.

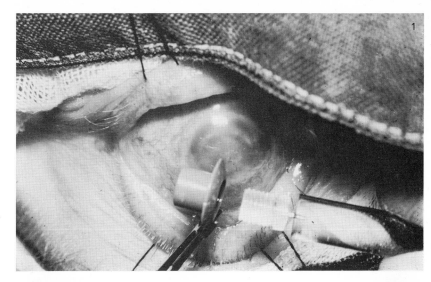

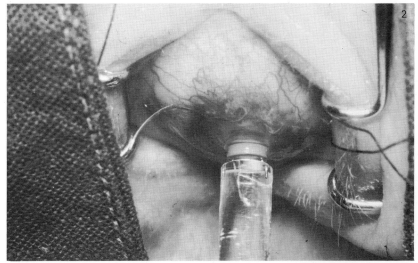

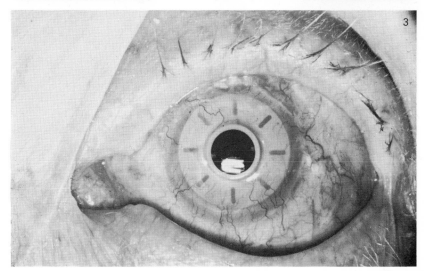

and tests are expected to go on over the next two years to explore other ways in which it can be used. So far volunteers have worn the lenses for up to 48 hours at a time. But they must be sterilised once a day – and therein lies the greatest difficulty in wearing them, according to the human guinea pigs. They're so comfortable, you can go to sleep forgetting you've got them in!

Although plastics technology has become of tremendous importance and has had widespread effects on the standard of living of most people in the world, it has created one disastrous problem: it is difficult to get rid of it. Its value lies in its virtual indestructibility: it won't burn, you can easily break it into small pieces or grind it into powder, and it won't rot, like organic materials. The vast majority of plastic refuse consists of food and drink containers, and it is a drink manufacturer that has helped develop a new plastic that solves this problem. The Coca-Cola company, together with Monsanto Chemicals, have put their soft drink on to the market in bottles that burn just like wood. The bottles, which are much lighter than their glass predecessors, are made of a new type of plastic that is impermeable to gases like oxygen and carbon dioxide, but since they burn they will not interfere with the operation of normal incinerators. Although Coca-Cola say it is too early yet to predict whether their production lines will convert totally to the new plastic bottles, they and other food manufacturers may be obliged to do so if anti-pollution laws become more stringent over the next few years – and the signs are that they will.

One of the few areas left relatively unpolluted by man in the last fifty years of industrial development is the Arctic. With the discovery of large quantities of oil in Alaska and Northern Canada, the attention of the ecologists has centred on this rare example of almost virgin country. They have caused major oil companies to rethink their pipeline routes, on the grounds that migratory behaviour of Arctic animals could very well be affected by the presence of the pipe, and the heat it would give out could alter the nature of the tundra over considerable distances. Now constructors and oil-men have discovered that they may be able to carry out their transportation, construction and drilling operations without ever coming into contact with the tundra. Their equipment need never even touch the ground, thanks to the latest development of the hover principle. It's called an Air Cushion Transporter, and consists of an all-steel box-like hull, with a flexible air containment skirt around its bottom edges, and a large-volume air-supply system mounted on its deck. Air is forced towards the bottom of the hull, causing the skirt to form a partial seal with the ground. As the skirt inflates, the hull lifts to its designated cushion height. Wheels, mounted on the bottom edges of the hull, allow the ACT to be towed along, carrying its heavy load of equipment, by a very much lighter vehicle, since the hovering 'trailer' weighs just enough to give the wheels directional purchase on the ground. The engineers have even installed a refrigeration system in ACT so that the blast of air will not melt the tundra permafrost. The version at present under construction measures 57 ft × 75 ft × 6 ft, and weighs, unlevitated, 100 tons. It will be capable of carrying 100 tons of equipment over shallow water, mud flats, tundra, and obstacles up to 4 ft in height. Air Cushion Equipment, the company that builds ACT, has already built hovertrailers on a smaller scale, which have successfully been used to move everything from pipe-laying systems to oil-storage tanks to stranded aeroplanes. The hovercraft principle may not have had the success once hoped for in the field of public transportation, but it is certainly moving into wider fields of industrial application.

Whatever the means for getting your industrial equipment from one place to another, the one thing you are almost always going to do when you get there is prepare to set it up. If it is to be housed in any kind of permanent structure you will need machine shops, and cutters to shape the hundreds of bits of metal needed during construction, no matter how much prefabrication is involved. The cutting lathes in those shops are about to undergo their greatest modification since they were first invented. If you look up at a neon light you'll be looking at the reason why. In that light the vapour is being excited into brightness by an electrical discharge. That's why it glows. What's happening is that at a certain temperature the molecular bonds of the gas break down, allowing the gas atoms to move about violently and at random, so that they collide with each other and release smaller particles: electrons. These in turn collide and generate radiant energy that causes the vapour to glow with increasing brilliance as the temperature rises. This cloud of matter in a perturbed state is referred to as 'plasma', and there is an increasing tendency to accept it as the next stage of matter, after solid,

Come fly with me. After a runway mishap this airliner straddled a metal fence, half in and half out of the airport. It rode back onto firm ground using one of the latest developments of the hover principle.
(See p. 171.)

Ten times faster with a plasma torch. This new machining technique uses a plasma torch to soften the metal that is to be removed.

liquid and gas. A plasma torch can be made by discharging electricity into a gas, and restricting the molecular collision that follows in a very small area by cooling the outside of the chamber where the activity takes place. At a temperature of about 11,000°C, the gas and the excited particles are released in a fine jet. This jet has already been used extensively for rough-cutting metal, or making holes in it at high speed. It has now been discovered by a British engineering research unit that by directing the intensely hot flame on to a small area of metal being cut in a lathe, near the cutting tool, the metal can be turned that much faster. The heated metal goes hot and plastic, making it very easy to cut. The heat generated is removed almost immediately with the metal itself and so tends not to spread and affect the rest of the metal mass to any significant extent. In fact, the friction between the cutting tool and the metal is now so much reduced that the usual liquid cooling and lubricating systems used in high-speed cutting are no longer necessary. What this all means to the man in the street is that if it is easier to cut hard metals which were previously not used because of the prohibitively expensive machine-lathes necessary, it is more likely that these metals will now be used in, and give longer life to, anything from a washing machine to a car gearbox.

The plasma state is the basis of another advanced technique for welding. It was developed at the British Atomic Energy Authority's Harwell Research Centre. In the Glow Discharge process, as it is called, electrons are generated by bombarding a solid metal cathode with ions from a gas plasma. These ions can then be directed to form a beam of virtually any shape: a ring, a straight line, or a very fine point to create heat. Its main advantage lies in the fact that the heating is absolutely uniform, so there is the minimum of residual stress round the weld. The beam can also be programmed to heat the weld area gently before switching to high intensity, or cool gradually after finishing the welding operation. With a system like this production-line techniques are possible, due to the ability to reproduce, to within very exact limits, the intensity, shape and temperature of the beam.

Another way of getting materials to bond is to stick them together with adhesives. There are many different types on the market, and with few exceptions a considerable amount of treatment of the surfaces is necessary if the adhesive is to be effective. Now a British company have developed a new adhesive, based on a chemical called cyanoacrylate. With a fixture time of seconds, this new adhesive will bond almost any industrial or domestic material – wood, ceramics, plastic, metal. The strength of the bond is considerable – in a steel-to-steel connection, the adhesive provides tensile strength resisting a pull of up to 5000 lb per square inch. The cyanoacrylic comes in three viscosities to be applied by hand or by special automatic equipment, and will resist continuous temperature changes between $-80°$ and $+80°C$. You can dip the bonded parts into alcohol or benzine for up to 24 hours and they won't come apart. The reason for this extraordinary strength lies in the molecular structure of the adhesive itself. Once exposed to the moisture in the air, the molecules bond together extremely quickly, causing the adhesive to harden in less than a minute. But it will react in this way only at the moment when the two surfaces to be joined are brought together. Unlike other adhesives, the shelf-life of this cyanoacrylic is considerable – up to a year, well after the point when most other similar adhesives have hardened and become unusable. This is thanks to a closely-guarded, secret stabilising agent which the developers add to the substance. There is only one drawback: don't get it on your skin. It's not poisonous, but, they say, you'll never get it off!

One adhesive manufacturer and a group of electronic engineers have got together to produce an entirely new way of building instant circuits. The system

Quick way to design and build prototype electronic circuits. Glue holds the parts in place until the design is perfected and permanent soldering begins.

Nodus — the metal joint that sticks without welding.

A completed roof structure is hoisted into place. One advantage with this system is that alterations and extensions are simply carried out.

is called Circuit-Stik, and with it you can put a series of connectors, conductors, transistors and other electrical components together like a jigsaw to find out if your particular design for a new circuit will work. The kit comes with 180 different parts, each piece backed with a pressure-sensitive adhesive and already pre-drilled, pre-plated, flux-coated, and ready for soldering, once you are sure your design will work. With the new system you can go from idea to finished circuit the same day, and if the design works when you run the current through it, you can begin manufacture plans within 48 hours of having the original idea.

The people involved with perhaps the largest-scale problems of joining things together are architects. For them the problem is not so much to find a way of making sure component parts of their structures stick together, but that they can take the *strain* of staying together. One of the latest developments in building design is particularly involved with this problem, dealing as it does in 'space frames'. These are structures using tubular steel. They are the subject of growing interest because they are aesthetically pleasing, they are the solution to many design requirements like large uninterrupted spans or potential extension in any direction, and they have a capacity for carrying a considerable number of overhead services. The lattice-work appearance of these 'space-frame' constructions is increasingly visible throughout the western world. But one of the factors that has always limited the constructors and designers is the kind of way in which one tube can be connected to another. There is no limit to the number of angles at which you can join tubes, as long as each individual join has its own connecting device, but what happens when a number of tubes converge on a common point? The answer is that they don't, if the designer can help it, because the problems involved have until now been considerable. Already experiments have shown that to develop a device that would permit the joining together of tubes coming from many different angles is impossible. The complex interaction of stresses would rip the join apart.

Now a device has been developed to suit at least the following requirements: that the joining tubes would be at right angles to each other; that the joint would be equal to the strength of the tubes coming together in combination; that the joint could be mass-produced; that no welding would be necessary; and that the structure would be competitively priced. The device is called the Nodus, and is expected to revolutionise the building of space frames. It consists of a central structure with four threaded openings into which the tubes screw. Beneath the central connection supporting bars lock on, and the whole structure is then tightened by the application of a bolt running vertically through the centre, holding a clamp which fits around the four tubes at the point where they join the Nodus. It is an extremely simple design, and variable in the sense that the threaded openings can be offset to take tubes from differing angles to the horizontal, provided that they are always at 90° to each other. Already this device is in operation in the construction of covered markets, warehouses, and other types of industrial buildings that require flexibility in construction and change in design from time to time.

One of the problems with the construction of high buildings is protecting the workers on the site from falls. This is particularly true for men working on vertical ladders. There are already safety-harness systems in use, but they are cumbersome and they tend to get in the way and limit movement. Moreover, you always need at least one hand free to move the belt up with you as you climb. The trouble is many spidermen and construction workers simply don't bother to use them. The result: hundreds of high-level men get killed every year, and thousands are seriously injured. It was in an effort to cut down on accidents by making a safety harness easier to use that led to one recent development by a British firm. It's called the Railok, and if you're wearing it you can climb or descend with both hands free. The safety secret lies in a rail that runs the entire length of the ladder in use, to which a device rather like a miniature roller-skate is bolted. On one side, in the rail guide track, there is a pair of tiny rollers. The other side of the device is attached to a safety belt by a metal connecting pin. As the worker climbs or descends, the rollers move the block along the guide-rail, as long as the pull is in the horizontal or upward direction. This allows the man on the job to climb using both hands, or descend slowly, or lean out from the ladder while working. If he slips and the block is subjected to a downward pull, it pivots vertically, lifting the rollers off the guide-rail and bringing four hardened-steel disc brake pads into operation, stopping the fall within inches. In early tests the brakes succeeded in stopping a load of 1200 lb, equivalent to the weight of six men, in the space of a few inches on a greasy rail.

The one basic need every development needs, whether it be in the field of construction, mining, aerospace or any of the myriad different areas of research going on in today's highly sophisticated industrial world, is power. By far the greatest source of power today is electricity, and although work goes on to develop some of the more advanced sources, like nuclear fusion, new techniques are also

Safety at height.

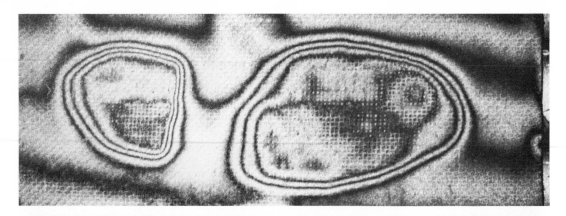

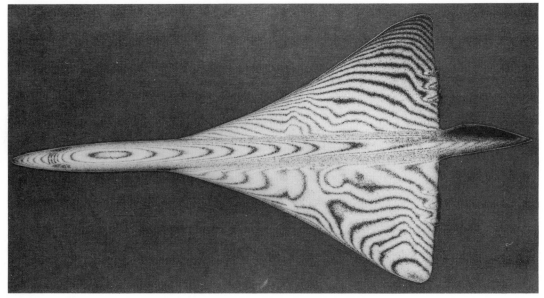

being explored at a more mundane, small-scale level. The latest micro-power-source is a new type of portable battery. It measures about 2 inches square, and will fit snugly into the palm of your hand. But it will produce five to eight times the power you could get from a conventional dry battery of the same size. The unusual thing about this new miniature power-source is that, in one sense, it draws power from the air. The battery is called the Zeta, and it works on the same principle of the electrolytic cell that powers the miner's warning device. The anode is held in an outer plastic case, and is formed of amalgamated zinc powder which also incorporates the negative terminal of the cell; it is in contact with the electrolyte, a concentrated solution of potassium hydroxide. The cathode consists of two layers: the outer is a micro-porous film which allows oxygen from the air to diffuse through it into the cell; on its inner side is a layer of catalyst, which turns the oxygen, diffusing through it to the electrolyte liquid, into hydroxyl ions. It is this transformation that generates the power. The current generated at the catalyst layer is collected on a fine wire mesh laid on top of it.

Obviously since it is the presence of air that activates the whole system, the batteries are packed in airtight containers, although at the moment the researchers are looking for a way to design some form of on/off 'air switch'. If they do, then it may not be long before getting power 'out of thin air' may be a distinct possibility.

Top The shoe that made its mark passed this way nearly 24 hours ago. But using lasers the imprint can be detected even on carpet.

Using the same technique microscopic changes in surface structure can be seen on this model aircraft after a wind tunnel test.

It is in the more specialist fields of research that this magical feeling about what technology can do is most evident.

Perhaps the most exciting of these very unusual developments is a technique that can, in one sense, show us the past. It is a way of detecting microscopic changes in the movement of certain materials which have been subjected to pressure. If you tread on a piece of carpet, for example, the fibres will appear to the naked eye to spring back into position almost immediately you take your foot away. The truth is that the fibres can take over 24 hours to complete that movement back to their original position. The scale of movement of course decreases considerably after the first few moments. The new technique allows an observer to see in detail the shape of the original object that pressed on the carpet up to 24 hours later. If you lay down on a wooden floor, the outline of your body could be clearly seen the next day. The secret lies in the use of the laser and its adaptation to differential hologram interferometry. A laser hologram gives a three-dimensional view of an object. If one hologram is taken after another at intervals of, say, 20 minutes, the minute changes in the surface under observation as it springs back into shape infinitesimally slowly register as a pattern. Superimposition of a series of the holograms will cause interference lines to show in those areas where there *has* been the slightest movement, down to 100,000th of an inch. Those lines will show because, in every other area but that where movement has occurred, every detail of the texture pattern will be the same, and therefore not show up as different from one picture to another through the series. It requires a relatively simple calculation to work out when the movement began, and therefore when the impression was originally made. The police are, understandably, the most interested party.

At the other end of the scale from this technique for showing movements too slow for the eye to see comes an American invention that enables you to see objects moving too fast for the eye to see. The inventor, a Mr Denaro of Concord, Mass., decided that the present stroboscopic systems were inadequate. These systems use a very strong flashing light which illuminates a rotating object once in every one of its revolutions. In this way the eye sees the object only in one position, and it therefore appears to be standing still. One limitation of this system is that the quality of the light does not permit detailed examination of the object. Another is that at the present state of stroboscopic art, only certain types of light can be used, and these do not cover the entire visible spectrum; so examination under infra-red or ultra-violet light, for example, is impossible. What Mr Denaro has done is to do away with the flashing light and replace it with an intermittent reflection. The idea is so extraordinarily simple that it's amazing no one thought of it before. The new stroboscopic device merely produces an intermittent *image* of the rotating object by rotating a mirror system in which it is reflected. The system consists of two orthogonally-mounted mirrors, rotating in the same plane and direction as the rotating object. When the rotational speed of the mirrors is exactly half the rotational speed of the object, you see a stable and stationary image. Since no special lights are needed, the image is in no way different from direct observation of the object when it is stationary. For the same reason, the new invention permits objects to be examined under differing kinds of light. Some stress effects only show up under polarised light, which can of course be used with the mirror system, thus making it possible to look for flaws in the design of high-speed rotational equipment in a way never possible before.

Of all the tools developed to provide man with extensions of himself, none has had a deeper impact on our industrial and social development than the computer. There are few aspects of our lives that are not affected in some way by it. The development of computers themselves has been spectacular since the

A strobe system that allows you to view movements normally too fast for the eyes to see. (See page 179.)

first one filled one entire floor of a building in 1947. Their size has been reduced dramatically thanks, in the initial stages, to transistorisation. The speed with which they calculate has increased almost as much, and the amount of data they can handle has increased by millions of times, as their storage medium changed from punch cards to magnetic tape to disc. Now it looks as if another 'computer revolution' is about to happen. Researchers at the Bell Telephone Laboratories have discovered strange magnetic properties in orthoferrites, which are materials composed of rare earth iron oxides. These oxides can be grown in crystal form, and when certain magnetic fields are applied to them, tiny bubbles form inside. These bubbles contain almost perfectly cylindrical magnetic 'domains'. The bubbles can be moved at high speed in the plane of a sheet of the crystal, by placing microscopic ferromagnetic antennae all over the surface of the material and attracting or repelling the bubbles. An alternative method of control is to apply a patterned magnetic field over the crystal. The advantage of this second method is that no wiring would be necessary. Now, since binary computers operate and store data based on the presence or absence of an electrical signal, known as a 'bit', these tiny bubbles can, by the presence or absence of their magnetic field, act as 'bits'. Since they are so small – a million of them fill one square inch of crystal – they represent a very high potential data density. Their manipulation can also provide the switching necessary for the processing and control of the data they carry. What these factors mean is that it looks as if the tiny bubbles are capable of 'being' the information and processing themselves at the same time – and all of the necessary activity can be contained within the limits of the crystal plate itself: a computer in a slice. The potential of this new discovery is almost unimaginably vast. It will almost certainly reduce the size and weight of complex computers to portable dimensions, and hastens the day when they will no longer be the costly and sophisticated machines that are beyond the means of the individual of today.

A microscope and television monitor are used to check a 'magnetic bubble' experiment. The section visible on the screen is about one-hundredth of an inch in diameter.

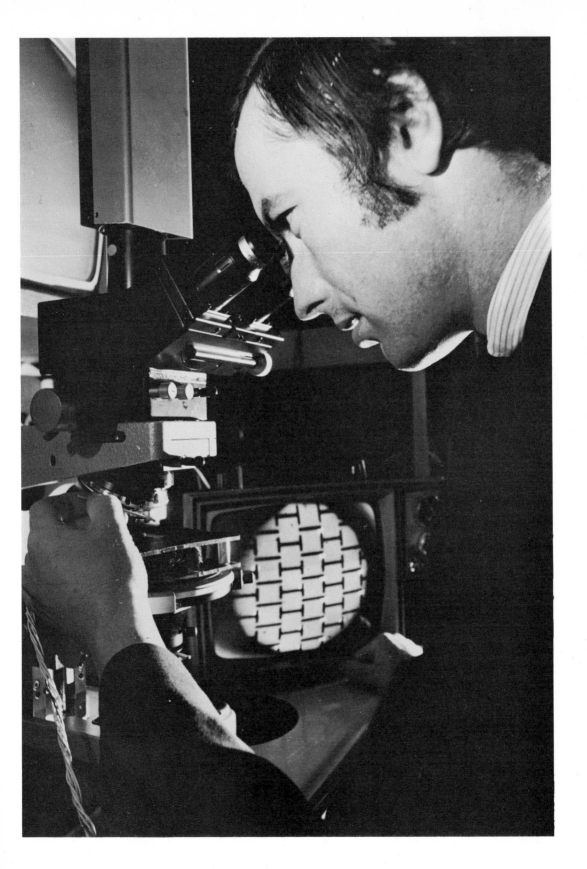

The operant conditioning of human brain waves immediately starts a lot of people worrying about brainwashing, thought control, and so on. The first advice that usually follows this reaction is to stop the research. But the psychologists engaged on this work point out the possibilities of using operant-conditioning techniques to obtain control over brain-wave patterns associated with epileptic seizures or insomnia. We may also be better able to resist attempts by others to control us if we have learned to obtain better voluntary control over our own brain processes and emotional states. In any case many scientists believe that, even if the dangerous possible consequences outweigh the beneficial ones, it would be a futile gesture to attempt to stop the research. Man, they say, is amenable to this type of psychological control, and this reality will not be changed by denying it. The attempt to become aware of the possibilities for the future, and to work to avoid the dangers, is better than trying to escape into ignorance.

Magnified magnetic bubbles (the bright discs) move through an electronic circuit. One bubble, somewhat elongated, can be seen in transition from one pole to the next. One day bubbles like these may be used as information carriers in computers.

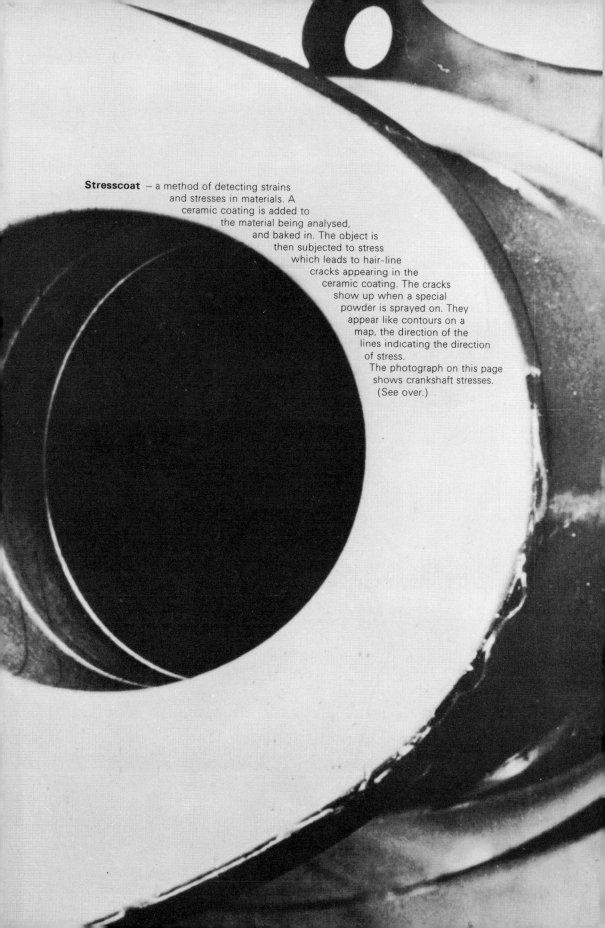

Stresscoat — a method of detecting strains and stresses in materials. A ceramic coating is added to the material being analysed, and baked in. The object is then subjected to stress which leads to hair-line cracks appearing in the ceramic coating. The cracks show up when a special powder is sprayed on. They appear like contours on a map, the direction of the lines indicating the direction of stress.

The photograph on this page shows crankshaft stresses. (See over.)

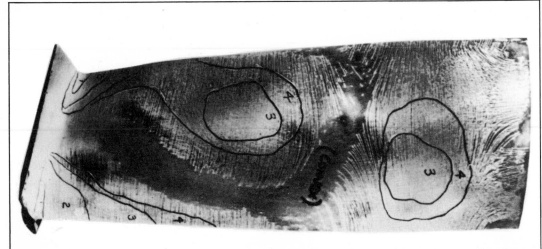

Engine inlet fan blade showing torsional vibration strains.

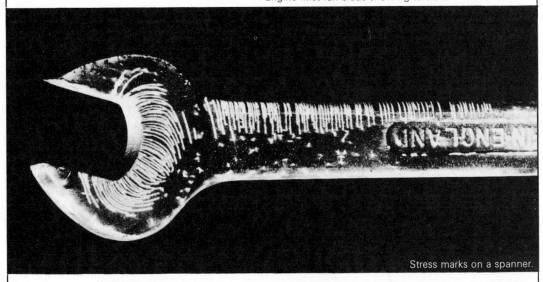

Stress marks on a spanner.

Winchester rifle breech stresses when fired.

Strain patterns on Rolls-Royce engine compressor blades.

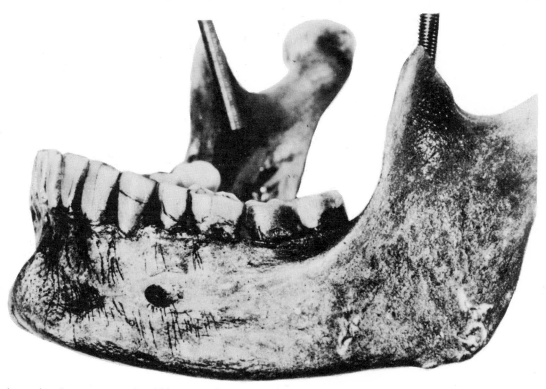

Lower jaw-bone stresses when biting an apple.

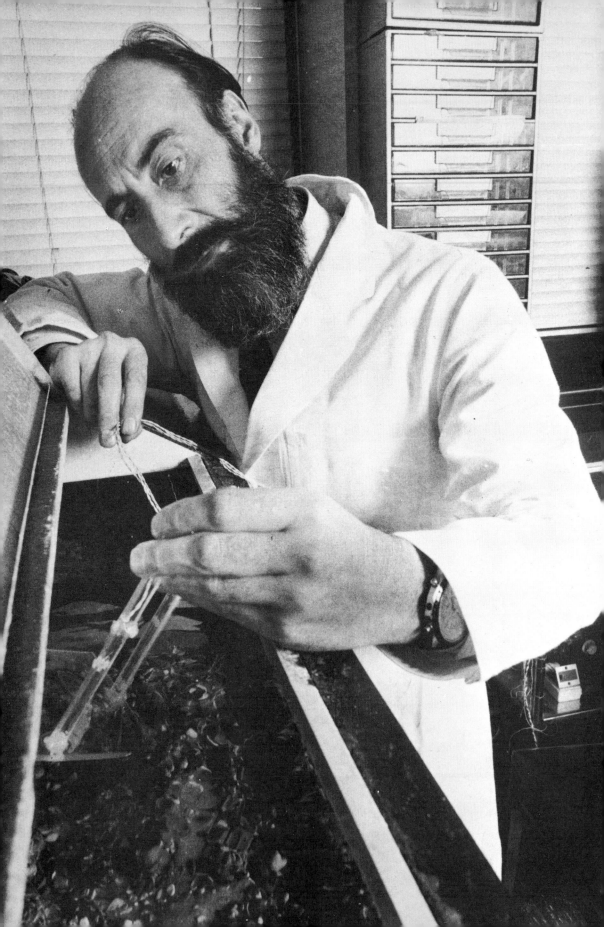

The Anatomy of Pleasure

Only in very recent years have scientists embarked upon the attempt to analyse the workings of the human mind and the way it governs our behaviour. It's not long since 'good behaviour' was ascribed to the influence of benign deities, 'bad' to the work of devils of one kind or another. When Sigmund Freud attributed major aspects of human behaviour to various manifestations of the sex impulse, it was for some people the most explosive theory since the idea that the world was round.

Elsewhere in this book we report recent striking advances in the understanding of the way the human mind works. Remarkably detailed discoveries have been made, such as the identification of specific biochemical substances in the brain which are closely concerned with behaviour and personality. But what is it that controls the pleasure centres in a human being?

The first steps that might lead to an answer were taken when the precise function of minute areas of the brain were discovered. In Britain the eminent neurophysiologist Professor Grey Walter was, by the mid 1940s, exploring the minute electrical currents generated within the brain itself by both conscious and subconscious thought processes. Once it had been established that specific types of electrical activity were associated with specific brain functions, the idea of some sort of induced 'feedback' immediately presented itself. If the source of electrical impulses generated in the brain could be identified, what would be the result of applying, from an exterior source, a similar electrical impulse to those areas?

During most of the 19th century the possibility of exciting the spinal cord and brainstem by other than physiological stimuli had been energetically debated in scientific circles. Many respected figures of the day categorically denied that it was possible to excite the brain in this way. However, in 1870 Fritch and Hitzig conducted experiments on anaesthetised dogs which demonstrated that electrical stimulation of one side of the fore-brain induced movements in the opposite side of the body. Further clinical studies by Paul Broca (1824–80) identified the position of those areas of the brain responsible for speech. It was this which led to wide interest in the possibility of discovering the functions of specific areas of the brain. As early as 1898 J. R. Ewald was pioneering the technique of introducing fine wires into selected points of live animal brains, and then studying the results of stimulation one or two days later when the animal had fully recovered from the operation. In a series of brilliant experiments which won him a Nobel prize in 1949, W. R. Hess demonstrated that automatic functions, such as posture, equilibrium, sleep, fear and rage could be influenced by the electrical stimulation of various cerebral structures. And for the first time it was shown that psychological behaviour like aggression could be induced by electrically stimulating the brain. But these discoveries did not significantly affect philosophical thinking of the day despite their inescapable and disturbing implications.

Even amongst biologists the implantation of electrodes in the brain attracted limited interest for 20 years. But in the 1950s the insertion of electrodes into brains became possible as a result of the use of micromanipulators and the creation of three-dimensional anatomical maps of the brain. The techniques of aseptic surgery, and the use of biologically inert materials, together with stainless

One way to stimulate a fish. Dr H. J. C. Campbell and the apparatus he has devised to observe pleasure responses in fish (see page 193).

M

steel anchorages, permitted long-term implantation of electrodes which could be left in the brains of experimental animals for several years.

It became a comparatively simple matter to implant electrodes into the brains of animals – and later people – without apparently damaging the efficiency of the brain tissue to which they were applied. The electrical impulses used were of very low energy, and the results obtained were so remarkable that a completely new field of research into the function of the brain was revealed. Since then the rate of progress has accelerated. For instance, transistors and 'solid state' devices have made a profound difference to the problems of connecting brain-implanted electrodes to external sources of power. An enraged bull has been stopped in mid-charge by the transmission by radio of minute impulses to an electrode in that section of the brain that was responsible for his aggressive behaviour. 'Free-range' hens have been induced in similar ways to peck on command, and a wide range of clearly recognisable behavioural responses have been induced in a wide variety of 'free' animals.

Needless to say such experiments have promoted grave disquiet amongst people rightly concerned about their validity, let alone cruelty. But the relevance of this work to the relief of human suffering has been demonstrated. In some cases of drug-resistant epilepsy, it was imperative to explore in depth the vital regions of the brain as a guide to possible surgical treatment. It also appeared likely that patients with intractable pain, anxiety neurosis, involuntary movements, and similar disabilities, could also benefit from cerebral exploration. The introduction of needles and catheters into the human brain was already a well-known pro-

People who use devices like this say it makes them 'feel good all over'. Now an eminent scientist is trying to find out why a fish and a pet crocodile appear to share the same opinion.

cedure used by neurosurgeons and, as the electrodes used for implantation were physically smaller, their introduction did not appear to involve any greater risks. Now hundreds of human patients in the United States, Britain, Japan and other countries have had electrodes implanted successfully in their brains for weeks and months. The patients have suffered no apparent ill-effects and have not been concerned by the presence of the wires in their heads. It now seems possible that the use of electrodes represents a far more desirable approach than the surgical destruction of portions of the brain, which had been regarded as necessary in the treatment of special cases of pain or involuntary movements.

It can be argued that therapeutic stimulation of the human brain with electrodes is comparable to, say, the use of cardiac pacemakers. As the pacemaker has kept defective hearts going, so paralysed limbs have been activated by programmed stimulators. Another example might be patients with permanent spinal block: urination has been induced by electrical activation of the bladder, thus avoiding the need for repeated catheterisation.

But the unique quality of cerebral stimulation by implanted electrodes is that it is not only bodily functions which can be induced, but also psychic activities such as memory and fear. In short, Man's thought and behaviour, so long the subjects of mystery and speculation, have been brought into the realm of scientific control, experiment and exploration, as a result of our newly acquired understanding of the neurophysiological mechanisms which control them.

Theories relating to all kinds of psychological and psychic phenomena may now for the first time be put to the test of controlled experiments and observation. Working under the auspices of the United States Public Health Service and the Office of Naval Research, Professor Jose M. R. Delgado, a pioneer and leading world figure in these studies, had reported by 1960 that electrical stimulation of the human brain had evoked a variety of mental phenomena, including vivid recollections of the past; sensations of fear and threats of unknown danger; feelings of pleasure and happiness, accompanied by laughter and humour; perception of words and phrases; and intuitions that the present had already been experienced in the past. Not only are these findings of considerable scientific interest which could lead to progress in the treatment of mental illness, but they also show that mental functions can be influenced by physical means – a fact of profound philosophical significance.

Current work is concerned with the possibilities of transferring inducted energy through the skin rather than directly to the brain tissue. This would make possible multi-channel stimulators so small that they could be permanently implanted underneath the scalp. Such instruments would be able to stimulate and command several predetermined areas of the brain. At the same time techniques of telemetry could be used to transmit back information about electrical activity within the brain of an individual, unrestrained by the gadgetry of an experimental environment. Professor Delgado has even invented a name for the miniaturised combination radio tranceiver and electrical stimulator which would be required. He calls it a Stimoceiver, and predicts that in the near future it could be used to control both bodily functions and emotions, i.e. to act as mental 'pacemakers' for people whose brains are incapable of performing certain forms of conscious or subconscious activity. He has gone so far as experimenting in providing sight and hearing to the blind and deaf by this method.

In the autumn of 1970 at Yale University, where he is professor of physiology, Delgado used a portable Stimoceiver on a chimpanzee called Paddy who was allowed to roam free with other animals. What made this experiment of particular significance was that the stimulating pulses were generated by a computer. A direct connection between a machine and the living brain of a 'free' animal had been created. The computer was programmed to recognise the pattern of electrical signals generated within the brain of the chimpanzee. It then sent back 'control' signals. The arrangement was effectively bypassing the senses usually needed for the mental processes of cause and effect. The computer was monitoring one part of the chimpanzee's brain so that it could tell another part what to do.

The implications of this experiment are enormous. Professor Delgado has achieved non-sensory communication – communicating with a brain, but by-passing the senses. He believes that this technique may be used to control a wide range of human conditions such as multiple sclerosis, Parkinson's disease, and even anxiety, fear, obsession and violent behaviour.

Such experiments have caused widespread concern about the moral and social implications of controlling the brain of the human individual. Delgado's reply is that it is not possible to *educate* by brain stimulation. All that can be done is to activate what is already there. He says that brain stimulation can be used to provide a window to the understanding of personality and behaviour, to improve, and compensate for defect or damage.

Elsewhere in this book (see Chapter 4) we report the suggestion from the United States that a government agency should be set up to prohibit lines of research which appear not to be in the public interest. Clearly the work of Delgado and his colleagues would be an early candidate for scrutiny by such a body. Yet if further research into the techniques of intra-cranial stimulation were to be prohibited, we might be deprived of potentially rewarding knowledge about the way in which the mind of man works, and about the development of techniques to aid the plight of people suffering from untreatable conditions ranging from blindness to sclerosis – to say nothing of the possibility of serious investigation of such subjects as extra-sensory perception, thought transference, or intuition.

Already his own work in this field has led to the advancement of a revolutionary theory by Dr H. J. C. Campbell of the Institute of Psychiatry in London. His particular interest focuses on what are known as the 'pleasure-sensing areas'. Dr Campbell is only too happy to demonstrate in his laboratory the eagerness of his experimental animals to accept the necessary preparations for the actual application of brain stimulation by electrodes. His rabbits will approach freely and put their heads into precisely the right position to accept the connecting plug to the minute socket implanted in the top of their skulls. As soon as the connection is made, the animal will make all speed to use its nose to press a lever mounted at the side of its cage. This triggers a minute electrical current which stimulates the 'pleasure centres' of the animal's brain. Some animals, including rats, have been known to respond in a quite extraordinary way: given the choice between a pleasure-exciting lever and one which produces food, rats have continued to operate the former until they have become exhausted by the physical effort involved. After a short period of rest, they have then resumed this self-stimulation of their pleasure senses. They will even accept known discomfort in pursuit of pleasure, such as crossing an electrified floor in order to reach what they know to be the pleasure-stimulating lever.

Dr Campbell is not prepared to commit himself on a precise definition of the nature of pleasure which his animals experience. In his own words, 'You could call it sexual pleasure, eating pleasure and drinking pleasure. It rather looks as if there is a fundamental thing that can be called pleasure, and that there are many ways of acquiring it – sexual activity being one, eating another. What is happening, we believe, is that the minute quantities of electricity reaching the animal's brain are having the same effect as any kind of activity which gives it a feeling of pleasure.'

In the United States mental patients receiving this type of treatment have made comments such as 'It makes me feel good'. Their behaviour has given every outward indication of increased pleasure – they have smiled, or laughed,

Monkey works the
light (see page 194).

become more relaxed and friendly. Even so Dr Campbell is not slow to admit
the possibilities of highly undesirable consequences were those techniques to be
abused. But it was not only because of these considerations that he became
dissatisfied by the whole process of inter-cranial self-stimulation by animals. It
seemed too far removed from the normal pleasure-seeking habits of animals in
the wild – although its similarity to Man's tendency to narcoticise himself is
inescapable.

Free, healthy animals, including Man, have to work hard for their pleasures.
Animals seem to get theirs simply by some activation of the sense organs – the
skin, the ear, the eye, the nostrils, etc. Dr Campbell argues from this that it
should be possible to get an animal to do something such as press a lever in order
to have his sense organs or 'peripheral receptors' stimulated. Unwilling to occupy
any of the serious research effort of his laboratory with further pursuit of this
private idea, he literally took his work home with him. He had an aquarium in
which he kept tropical fish. Into their water he inserted an electrode arranged in
such a way that, as the fish swam across a beam of light, a small amount of
electricity would pass between the wires and 'across' the fish. The fish immediately
began to swim deliberately through the beam of light in order to receive the self-
stimulation of the electrical current. Campbell likens this sensation for the fish
to that of a finger being gently drawn across the skin. He believes it to represent
the origin of the human stimulation of kissing, fondling and stroking. Most
animals like to be stroked, and 'fish tickling' is a long-established art.

The known simplicity of the fish's brain heightened Campbell's interest in
the subject. He went on to investigate the effect on reptiles and amphibians. A
pet crocodile, normally quite content to remain in one position for hours on end,
walked backwards and forwards between electrodes in order to enjoy the
stimulation of his scaly hide.

Since fish, reptiles and amphibians are early products of the evolutionary
scheme, Campbell arrived at his extraordinary theory. He suggests that the
appetite for peripheral self-stimulation is the very first piece of neuronal organisation. 'When brains developed to the level of the fish,' he says, 'which is again
rather high compared with that of, say, a worm, built into that ancestral brain-
computer was a command – activate the pleasure areas.' He applies his theory

to explain the extinction of some species, the evolution of others, and the evolutionary organisation of various parts of the nervous system. 'One can look at the whole of evolution,' he says, 'as the body getting better and better at seeking stimulation of the peripheral receptors; of getting a wider range of better receptors such as those in the eye; and a muscular system developed to enable the individual to achieve this pleasure more readily.'

Even crocodiles do it: Every time this pet croc swims between the plates a small electric charge makes its hide tingle (see page 193).

In order to explore the phenomena of peripheral stimulation in primates, Dr Campbell has experimented with the small and friendly squirrel monkeys of Peru. Given the capability of switching on a 500-watt bulb by pressing a metal rod, the monkeys quickly respond. By excluding the heat of the bulb from the monkey, it was possible to demonstrate that it was the bright light only which attracted it. With the aid of an automatic counter, it was established that during the first day of his experience, one of the monkeys pressed the rod some 16 times. After 3 or 4 weeks he was switching on the light as much as 500 times in 15 minutes. Similar results were achieved with five other monkeys.

But unlike the animals given the opportunity of direct self-stimulation via electrodes planted in the pleasure-sensing areas of their brains, animals exposed to peripheral self-stimulation behave in a far more natural way. Dr Campbell cites the example of the almost irresistible attraction of a bowl of strawberries. After a comparatively few mouthfuls one tends not to want them any more. He suggests that this is entirely analagous to that of the monkey with his bright light. Unfortunately for him, however, the monkey has only got the light with which to give himself pleasure. In a wild state, after sating himself in one way he would turn to something else. After eating a completely satisfying meal, an animal in the wild does not stop behaving. He may go to sleep, and thereafter he will set off to seek pleasure in a different manner, i.e. to stimulate different peripheral receptors. This has led to Dr Campbell's new hypothesis of behaviour, which he suggests produces a unification and a possible explanation of its entirety.

It is possible to delineate certain areas of the brain as receptors – the function of which is to pass on nerve impulses from sensing organs to the pleasure areas. It is also possible to delineate the connection between pleasure areas and the motor or muscular system. When an animal presses a lever, he transmits an impulse directly to the pleasure areas of his brain. But this 'bypassing' process is unnatural. When an animal is eating, his taste buds are being stimulated. These impulses are passed up through the nervous system to activate the pleasure areas of his brain. But after a while the taste buds do not respond in the same way. It is a basic property of all receptors that after a while, even when the stimulus continues unchanged, the receptor ceases to respond. Therefore, after we have eaten sufficient strawberries, although their taste has not altered, the pattern of impulses changes, and it is this pattern which determines when we have had enough. But there is built into us a drive to keep the pleasure areas activated, so we scan our environment in search of other sources of pleasure. Dr Campbell suggests that this hypothesis can be applied to account for the whole of evolution and all behaviour with the exception of abstract thought. 'A basic requirement of being alive,' he says, 'is to have the pleasure areas activated.'

Not all scientists, let alone philosophers, are prepared to accept the Campbell hypothesis, which he admits requires a great deal more experimental work before it can be substantiated. However, it does account for many of the anomalies of, for example, Freudian theory. It puts on precisely the same plain the importance of pleasure derived from sex, listening to Mozart, or eating a bowl of strawberries. All these are simply different types of stimuli for pleasure-seeking responses. Hillary and Tenzing, for instance, climbed Everest in pursuit of that basic objective, rather than for any deeply hidden sexual requirement. Although sex may be a very effective way of producing pleasure, it is not fundamentally different from dancing, or painting a picture. In answer to the question 'Where does that leave Freud?', Dr Campbell replies, 'Well, it doesn't leave him, it brings him in rather better.' Some of the criticisms that were aimed at him can now be removed. Freud's point was that on some occasions (*sic*) just about every form of activity had a sexual basis. Now according to Dr Campbell, we can see that what he really meant, although he didn't know it, was that all forms of activity have a pleasure-seeking basis. Because sex happens to be an extremely effective one, he likened a sunset or pipe-smoking to sexual activity. In fact they are all simple pleasure-producing activities.

To Dr Campbell, this hypothesis is of much more than purely academic interest. Its acceptance, and we are ourselves much attracted to it, makes large elements of psychology and psychiatry more understandable (see Chapter 9). There are cases of mental derangement for instance, in which the patient ceases to behave at all – endogenous depression in which the individual withdraws and simply sits unmoving in a corner, without speaking, apparently without seeing, and helpless to fend for himself. What would be the effect of the insertion of an electrode into the pleasure areas of the brain of such an individual, and a button he could press which would at least bring him back into the world? It might even be possible to do this by some form of peripheral self-stimulation. A further potential for treatment might be the condition which sometimes reduces young girls to the point of death by starvation because they lose all interest in eating. An increased sensory input from their taste buds might induce them to eat again normally.

The technique could even be relevant to delinquency. A few years ago a group of American psychiatrists suggested that at least some forms of adolescent

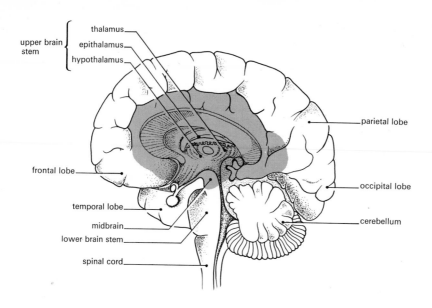

upper brain stem {
thalamus
epithalamus
hypothalamus

parietal lobe

frontal lobe

occipital lobe

temporal lobe

midbrain

lower brain stem

cerebellum

spinal cord

During its evolution the human brain developed two paired hemispheres. The lower ones formed the cerebellum which controls balance and integrates body movement. The upper pair, called the limbic brain, includes areas like the cingulum, hippo-campus, thalamus, epithalamus, basal ganglia, midbrain, and amygdala. This is the centre of emotions in the human brain — the part that contains the pleasure-sensing areas.

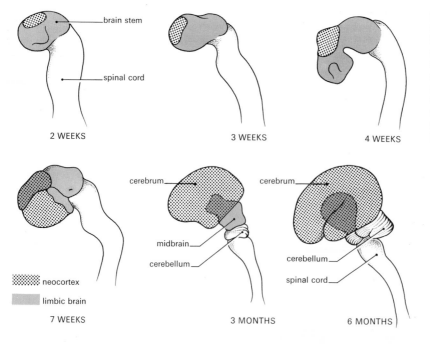

brain stem

spinal cord

2 WEEKS

3 WEEKS

4 WEEKS

cerebrum

cerebrum

midbrain

cerebellum

cerebellum

spinal cord

neocortex

limbic brain

7 WEEKS

3 MONTHS

6 MONTHS

Prenatal stages in the development of the human brain. As it grows the human embryo goes through stages roughly resembling lower animal forms e.g. fish, amphibian, reptiles, and then mammals (see page 193).

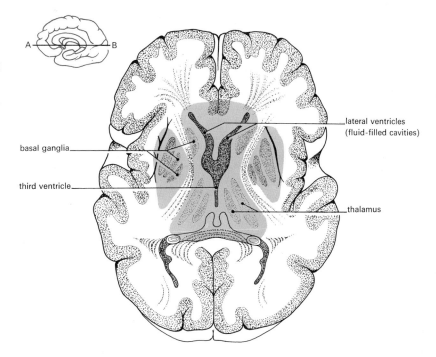

If you could look from above into the centre of a human brain, the areas around the lateral ventricles would be the limbic brain. It's the part that controls all emotions – including pleasure.

basal ganglia

third ventricle

lateral ventricles (fluid-filled cavities)

thalamus

delinquency were due to a fault in the thalamus – an area within the brain. The individuals concerned did not appear to derive sufficient stimulation from 'normal' behaviour. In other words they were not getting sufficient pleasure, so they had to go out and do things which brought them into conflict with society and the Law. Dr Campbell does not suggest that in such cases a flashing light, or electrodes like those which stimulate the scales of swimming fish, would be sufficient. But he does believe that medical engineers might be able to invent a new device for stimulating the peripheral receptors in delinquents so as to meet their specific requirements for pleasure, and in so doing offset their frustration.

But within the accountability for behaviour of Campbell's hypothesis there remains the stumbling block of philosophy, and here he admits defeat. Whereas every other form of activity could be described as animal (even in listening to music, or watching television, we are doing what the fish, the crocodile, and the monkey were doing but, thanks to technology, in a more efficient way), philosophy remains uniquely a human capability. Dr Campbell says, 'It is the only thing that I have not yet been able to incorporate into this theory. I'm still working on it and perhaps I will be able to.'

Philosophy apart, it is not difficult for any normal human being to accept the importance to his well-being of the stimulation of his sensory receptors. They are the sole reporters to the individual of his environment. But their powers of response and communication generate their own special language. The exchange of glances between lovers may be difficult to define even in the clinical terms related to the work of Delgado, Campbell and their colleagues. But to the principals involved it is more eloquent and meaningful than anything attainable in conventional language. Dr Alex Comfort, Director of Research in Gerontology at the Department of Zoology, University College London, suggests that as well as the language of the eyes, there is the language of noses. Is there, he asks, a subliminal language of odours between humans which excites fear, hostility, sexual attraction and a wide range of comparable responses? It is a language of such low intensity that it is not difficult deliberately to suppress it. All but the sharpest-nosed may not be consciously aware of its existence. But the contemporary demands of fashion and social behaviour may, in the view of this distinguished scientist, be responsible for potentially disastrous misunderstandings.

His theory centres about substances called pheromones. They are substances secreted by one individual which, by their smell, affect the behaviour of another. They control the behaviour of ants and much insect-mating, and are also widespread in mammals. Some mammalian odours are straightforward signals. The labelling of territory with urine is an example. The action of a true pheromone, however, is more direct. While it is a signal, its action is more like that of a hormone, and many pheromones in mammals have a chemical shape rather like a steroid molecule, and might have been derived from one.

Examples of the straightforward signal-type of pheromones are well known – the attraction of a bitch in season, or the rejection by a ewe of a lamb which is not her own. But the true extent of the pheromone influences in mammals only began to appear in 1961 when it was discovered that the odour of a strange male can terminate the pregnancy of mice. More recently it has been discovered that putting female monkeys on the 'pill' inhibits the sexual behaviour of the male. The communication seems to be by odour.

There is no evidence that pheromones do not operate in humans, and indeed there is some evidence that they do. Many psychiatrists believe that odour is among the psychological 'cues' which operate in certain situations, and schizophrenic patients who are hypersensitive to external influences claim to be able to 'smell' hostility. Some psychiatrists have themselves even claimed to be able to 'smell' schizophrenia. This possibility has subsequently been borne out by the identification of the substance involved as trans-3-methylhexanoic acid (see Chapter 9).

No one would question the psycho-sexual importance of human odour. Large-ring ketones such as civet and musk, which are mammal sex odours, have been for centuries ingredients of My Lady's perfume. But Dr Comfort suggests that our culture has been too squeamish about normal healthy body odours. We use deodorants, toilet water, after-shave lotions, and the like, at our peril.

All this would be pure conjecture but for a series of remarkable experiments and observations made recently, which suggest an important role for human pheromones in the formulation of relationships, response attitudes, and even the timing of cyclic bodily functions.

The belief that developed organs such as the appendix and tonsils 'do nothing' and can be removed without physiological effect is a naive one belonging to the last century. As antiquated may be the view that certain glands associated perhaps with the secretion of pheromones may have outlived their importance to contemporary Man. The time may be at hand, says Dr Comfort, when deodorants, intimate and otherwise, will be regarded as nothing less than environmental pollution.

Meantime the development and production of perfumes calculated to attract has grown into a major world-wide scientific industry. It is based, of course, on the premise that if the article smells nice – the smell gives pleasure – it is more likely to induce in the potential customer the response which will make him reach for a cheque-book. Now new, and sometimes commercial, methods of stimulating the pleasure receptors, which combine art and technology, are being rapidly developed; and it's hardly surprising to learn that a role has been found for computers in Man's quest for pleasure. Computer-formulated dance routines have been performed by several of the leading ballet companies in the West. Computers have been programmed to write their own special form of poetry.

Whether the artist is under an obligation to stimulate the pleasure areas of

A roomful of melody —
and it's called the
Musys. It is one of the
latest devices for
producing elec-
tronic music. At its
transistorised heart are
two computers, a fast
paper-tape reader/
punch and a magnetic
tape drive.

A computer music
score. It is a new work
entitled 'Probabilistic
Selection No. 3.'
Only Musys can play
it.

```
PROBABILISTIC SELECTION NO. 3.

SWEPT PITCH, FILTERING AND DYNAMICS WITH OCCASIONAL SURPRISES

08 -> A5 -> F2 -> AMPEX 2
11 -> O8(FREQUENCY CONTROL)
12 -> A5
13 -> F2(FREQUENCY CONTROL)

PERFORMANCE RATE=7

"

M=72 "SET DURATION"

"! BUS " 1\! "CONTAINS WAVETEK DATA"

Y=0 Z=0 100(M-Y[ "LOOP UNTIL Y>M"
A=3↑+5 B=6↑+2 C=6↑+8 D=15-A E=#RAND ;
A(#WVTK E; E1.15. A4.D. D=D+1 #WAIT B; E1.4. #WAIT C;
10↑-9[#SURPRISE ; "OCCASIONALLY"] )
#PROB X,2,#L1 ;,#L2 ;; #WVTK X;
E1.15. A4.3↑+12.
#PROB X,4,#L3 ;;#L4 ;; #WAIT B*X; E1.4. #WAIT C*X;

H=20↑+4 "NUMBER OF NOTES"
H(#WVTK #RAND ;; E1.4↑+11. I=10↑+4 #WAIT I; A4.4↑+11.
     E1.5↑+2. I=16↑+4 #WAIT I;)])

"CONTROL THREE INTEGRATORS"

B=3 3(B=B+1 1"! BUS " B\! Y=0 Z=0 100(M-Y[B+23.63↑. #WAIT 40↑+20;]))

$ "MACROS"

WVTK D1.0. O8.0. %A; D1.1. @ "WAVETEK (O8) CONTROL"
WAIT T=%A<63 T1.T. #COUNT T; @ "WAIT AND MEASURE ELAPSED TIME"
COUNT Z=Z+%A Z-100[Y=Y+1 Z=Z-100] @ "SECS & HUNDREDTHS COUNT"
"PROBABILISTIC LIST SELECTION MACRO"
PROB R=100↑ X=0 N=1 %B(R-X[X=N%C+X] N=N+1) N=N-1 %A=N%D @
RAND 2047↑&1023+1024@ "WAVETEK NOTE GENERATION"
SURPRISE Q=0 20↑+20(#WVTK Q; #WAIT 5↑+5; Q=Q+50↑) @

"PROBABILITY SELECTION LISTS"
```

any brain other than his own is a matter of classical debate. There can be no doubt, however, that a large proportion of contemporary artists appear to be fascinated by the possibilities of achieving unusual effects by the use of unconventional stimuli, or conventional stimuli used in unconventional manner. It is not enough to excite the eye or the ear alone or in combination. The 'feelies' of Aldous Huxley's imagination are now with us as a result of different concepts amongst a wide variety of artists. The sculptor has long been interested in texture, not only because of appearance and form but also because of its 'touch' impact. The different appeals of smooth and rough surfaces have been exploited in sculpture as old as the art itself. But many contemporary artists have created pieces intended to achieve their effect by touch as much as by sight.

A 'feelie box'. It's intended for two people. Hugh Davies who created it says they could hold hands in the middle 'where there's soft fur all round'. Normally the 'feelie box' would be closed with just slits at either end to get your hands in.

The texture of plastic has also excited many artists, and the development of these materials has led to a completely new generation of art-form shapes in the design of inflatables of one sort and another from furniture to theatres. The Event Structure Research Group in Holland has won itself widespread notice in this field. They constructed a pneumatic bridge spanning the Maschsee in Hanover for the Street Art programme there in the summer of 1970. It consisted of a tube 250 yards long and 10 ft in diameter, made mainly from transparent PVC with a floor panel of polyester fibre. The floor panel also incorporated a water compartment to prevent the tube from twisting and the whole thing was ballasted every 12 ft to concrete blocks. It was kept inflated at very low pressure by one centrifugal blower. The Group has plans for many potential further applications of their design, not only as works of art, but also to perform entirely commercial functions such as over-water goods conveyors, ship to shore loading, and so on.

Another of their creations was called 'Cloud'. This was made from white PVC sheeting, and was intended to be suspended or flown with helium gas. It was first shown at the Stedelijk Museum in Amsterdam, and its effect was supplemented by the amplified sound of wind and rain, and the projection upon its surface of images of sky and clouds. The Event Structure Group hope to use this creation to produce full-scale manifestations of natural weather, but incorporating their own improvements – coloured rain, sweet-tasting snow, and multicoloured suns.

A group of artists in Britain have produced a project known as 'Pavilions in the Parks'. These consisted largely of plastic, inflatable tent-like structures, the purpose of which was to create an environment which would not only be protective but also complimentary to the wide range of artistic objects displayed in them.

Another British group, 'Action Space', describes itself as a flexible association of artists concerned with exploring the area of activity where the arts overlap play. Under the guise of play they aim to introduce children and whole communities to a more active and participational association with the arts. Action Space uses frame, tent and pneumatic structures, movement, lights, slopes, sound and drama to develop the visual, tactile and space 'awareness' of people. The group sets out to create 'total participation environments'. They have worked on community play projects, arts festivals in schools and colleges, and in the London region they have specialised in areas of major socio-environmental problems such as Wapping, Notting Hill and Camden.

In some ways more conventional in their approach, but equally preoccupied by the pursuit of totality in the experiences which they offer, is another British team called 'Archigram'. This is a group of architects who have designed some extremely interesting and successful architectural ideas incorporating the concept of 'total environments'. They were regarded as something of a joke in the more conservative circles of their profession until 1970, when they won a highly competitive international competition for the building of a multi-purpose events

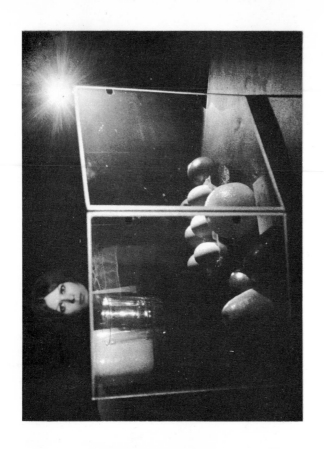

Hold the page in front of you and cross your eyes. A third photograph will begin to appear in the space between them. Concentrate attention on this third image and it will gradually fall into focus and appear three-dimensional. The face in the background is Margaret Benyon who thought up the idea.

facility in Monte Carlo. Theirs was the only design submitted which did not involve structures built above ground. Instead they proposed to dig a 250-ft-diameter hole containing highly mobile, highly versatile modules which offered the site the capability of housing, as necessary, a circus one night and, say, a chamber orchestra the next. Another of their projects is called 'Instant City'. This concept is still very much in the formative stage, but the latest idea is to float various structures into selected areas by barrage balloon, or by employing lighter-than-air gases in inflatable shapes.

A good deal more bizarre in his approach to the sensory impact of his work is Juan Christo, an artist who specialises in wrapping things up in polyurethane. By 'things' he means almost any type or scale of structure from a medium-sized skyscraper to a chunk of rocky coastline or a mountain canyon. In comparison, the creations of Margaret Benyon, of the Department of Fine Art at Nottingham University, are extremely conventional. She has made a considerable impact with her hologram photographs (the creation of third-dimensional effects using laser light and double or triple exposures sometimes involving mirror inversions). Although Miss Benyon is alone in her field in this country, there are several artists working along similar lines in the United States.

From this it will be seen that there is a large and growing body of artists and art groups who are, almost certainly unconsciously, providing supporting evidence towards the credibility of many of Dr Campbell's ideas concerning sensory reponse, particularly in view of the widespread enthusiasm with which many of their projects are received, and noticeably by uninhibited youth.

Much less subtle in its approach, and indeed highly analagous to the stimulant electrodes which Dr Campbell placed in his aquarium for the pleasure of his fish, is a device marketed under the name of 'Slendertone'. The manufacturers claim that its principal purpose is to serve as a slimming machine, and that under a doctor's supervision it can also be used for rheumatic and articular conditions, to increase circulation, and other therapeutic applications. The standard home model consists of a '4-outlet battery-operated unit with 8 body pads'. The wet pads are attached to those areas of the body where stimulation is required. It works on the long-established principles of 'Faradism' which has been used in hospitals for many years to re-educate muscles. Mild electric shocks engendered through the wet pads produce a galvanic effect upon adjacent muscles. The users describe the resultant sensation as 'pleasing', 'relaxing', 'it makes me feel good all over'.

And where, we ask ourselves, have we heard that before?

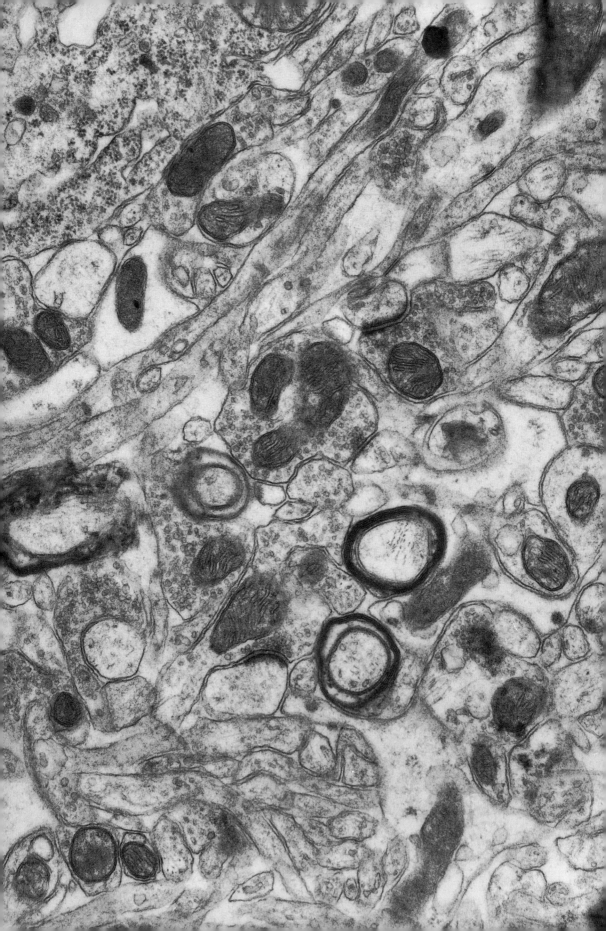

9   All in the Mind

At the corner of University Boulevard and Fifteenth in San Diego, California, stands a building that still looks outwardly like the bank it once was. But the presiding genius over its current activity bears no resemblance to his predecessors in the panelled central office. A 6 ft 1 inch, explosive figure with bristling black beard, whose energy threatens to burst his jeans, and who talks at 125 words per minute. In conversation he says things like: 'Psychoanalysis as a treatment is a load of crap. If it ever did anyone any good it was because the consultant in that particular case was a Houdini. Only an escapologist could break the fetters of conventional analysis sufficiently to be able to help his patient.'

Dr Tom Rusk, a Canadian Jew in his early thirties, roars with laughter when he explains to his friends that he left McGill, after graduating brilliantly, because of anti-Semitic prejudice in the medical school. As may be imagined he is a controversial figure amongst the more conservative elements of public health and university administrators even on the West Coast of the United States. At the same time his attitudes are in many respects typical of the radically independent views and ideas which make the University of California the exciting place it is.

Rusk is a psychiatrist whose approach is as adventurous and exploratory as that of any of the biochemists working on their advanced molecular theories in the laboratory buildings a few miles away. The former bank in which he and his small group of colleagues work is a free, 24-hour psychiatric clinic, available to anyone in need of its services. In a sense there is no other way in which it could be operated because Dr Rusk's speciality is Crisis, and his theme is response to urgency. In his view most of the new community mental health centres have proved to be little more than hospitals moved into town, with emphasis on the care of the in-patient, individual treatment, heavy use of drugs, and early discharge with little or no follow-up procedure. Some centres have open doors and are prepared to be innovatory, but they divert the 'acutely disorganised', the seriously sick and the aged to the major hospitals.

The approach of his San Diego clinic is radically different. It focuses on immediate attention to patients in a state of crisis, practises 'crisis group-therapy', and provides continuity, not only in time but also in terms of the people involved in the treatment of any individual case. A direct and personal contact is established as soon as possible between the patient and one senior member of the staff – a primary contact in every sense. The clinic provides intensive day care and can cope with brief overnight stays for acute cases. Long-term follow-up groups, social advice and the development of community resources are all key parts of its function. At all levels the professional and semi-professional staff work together as a team as closely as possible, while each retains his own specialist interest and responsibilities.

But the San Diego clinic is a research organisation as well as a mental health centre, and consequently the objective assessment and evaluation of its own function is accepted as one of its fundamental tasks. Its purpose is to meet the urgent needs of the present and future. Nowadays stress is accepted as being an inescapable hazard of modern life. The psychological definition of a state of crisis is when 'a person faces an obstacle to important life-goals that is for a time insurmountable through the utilisation of customary methods of problem-

Part of two brain cells magnified 20,000 times. The membrane that separates them can be seen in the top left-hand portion of the picture. The incredible complexity of a single cell is a measure of the difficulty scientists face when they try to unravel the mysteries of the brain itself.

solving'. In lay terms, 'you can't cope any more'. There must be very few people who have not found themselves in this condition at some time or another. Normally we can accept it as being a part of life – certainly a part of growing up. Normally we can meet the challenge ourselves, though probably with the assistance of relatives and friends. We pull ourselves together, and get on with the business of living. But in fact the crisis state can represent a very serious psychological condition. Expressed in the professional language of the psychiatrist, this is when the ego's adaptive and creative capacities are inadequate to handle the stimulus. The mild anxiety evoked by any new stimulus is replaced by increasing anxiety which constitutes a persistent and increasing threat to ego-equilibrium and integrity. This equilibrium is essential to what we consider to be normal, reasonable behaviour. Human beings have three fundamental needs – affection, security and significance. The satisfaction of all three assures one's sense of well-being. Parents, especially mothers, originally provide them, but the threatened loss of any of these requirements can have the same psychological effect as the actual loss of a parent in very early life, and is thus a threat to one's very existence. According to this theory the loss of a job, or even the *threat* of becoming unemployed, could psychologically represent a small child's loss of its mother.

Rusk lists the readily observable symptoms of a person in crisis – insomnia, excessive activity, fatigue, restlessness, muscle tension, and focused attention. Often the individual concentrates his thoughts on a single, irrelevant, inappropriate aspect of his immediate experience. A common example is concentration upon some aspect of the bodily ills normally associated with anxiety, such as worry about a heart attack or stomach ulcers, while at the same time the victim sees his whole life as being in a state of catastrophe. The overwhelmed ego, unable to cope, tries to escape, becomes anxious and attempts to recapture the previously satisfying behaviour and attitudes of childhood. All needs are felt as intense and imperative and the sufferer is impatient for their gratification. A true assessment of personal ability, the differentiation between what is real and what is imaginary, and the capability for mature thought decrease. Unfortunately a product of all this confusion is that the sufferer's behaviour alienates the sympathy and support which he most needs from his close relatives and friends. They tend to 'put off' the people whom they most need. Lonely, confused, worried and desperately unhappy, they are in urgent need of someone to help them, and are aware of this. Consequently, and this is the core of Rusk's theory, at the very height of the crisis they are in an almost ideal situation to achieve benefit from the attentions of a skilled psychiatrist. If the therapist is willing to break in as the omnipotent figure the patient craves, the anxiety will diminish, the regression will reverse, and the normal 'ego functions' will return.

All this will make a great deal of sense to the responsible and experienced parent, and indeed to anyone who has been in a position of responsibility or leadership over other people. But analysed and expressed in these terms, it will be seen to dictate the entire manner in which the San Diego clinic is conducted and to underline its differences with the more conventional approach to established psychiatric practice. However, in attempting to 'strike while the iron is hot', as opposed to postponing activity until the situation has cooled down, presents a number of disadvantages. 'Why don't you go home and have a good night's sleep and we'll talk about it in the morning?' is precisely the opposite of the Rusk approach. But confronted with a total stranger, clearly in a very disturbed state, he and his colleagues have to make intelligent decisions on the basis of very incomplete information. They do not have time to make good those deficiencies. The patients themselves are likely to be extremely unco–operative in many ways, and it can be difficult to arrange the necessary follow-up.

What could be described as 'scenes' are by no means uncommon in the clinic.

The poignant tragedy of mental illness. A 15-year-old Japanese girl who behaves like a baby. She cannot stand up, and if left alone shrieks, tears rolling down her cheeks.

Tom Rusk is certainly not one who hesitates to shout back at a patient who may shout at him, if he thinks that this will help. Conversely he will not hesitate to *show* how much he cares about the patient's predicament. He will hold hands and comfort as best he can. He believes that to help he must be seen to become truly involved, to understand, and to care. Support of this type is desperately desired and yet, in his experience, so rare in the lives of most people by the time they are in crisis that the therapist providing it discovers that he has all the 'leverage' he needs to accomplish his short-term objectives.

But ultimately, as in all psychotherapy, the therapist works towards the patient's resumption of as much responsibility for his own life as his current and potential capabilities will allow. This is achieved by the follow-up procedures which include group therapy aided by as many technological tricks and devices as can be brought to bear, including television cameras and the playback of video-tape recordings. People in acute emotional distress have an aversion to mirrors. Rusk says he has had patients who, asked to pretend that his hand was a mirror, were unable to look at it. But the fascination of seeing oneself on television is a phenomenon all too familiar.

The Rusk team attaches great importance to family influence. Part of their follow-up procedure is the establishment of a relationship with the patient's family as a whole. In the absence of family or friends, they are prepared to undertake the role themselves, accepting the consequent drain on their own emotional resources. Tom Rusk himself carries a bleeper – a hip-pocket walkie-talkie – when he is away from office and home. His concept of the 'ever-open door' means for him a 24-hour involvement with the people who have called upon him in their hour of crisis.

Dr Rusk is fully familiar with the problems of drug abuse both within and without the medical profession, but his definition of abuse may not in this, as in many other matters, coincide with the majority view. He is at pains, of course, to differentiate between the habit-forming and non-addictive drugs, but he believes it possible that Man may have a basic requirement to intoxicate, if not narcoticise himself. History bears him out.

Throughout recorded history and in almost every culture people have taken chemical substances to change their mood, perception and thought processes. The earliest records suggest that this started in North-eastern Europe and has existed since 2000 BC. The magic mushrooms of the Aryan invaders of India contained the active ingredients of LSD. Papyrus documents indicate the use of wine by the Egyptians about 1500 BC. The opium poppy appears in records as early as 1000 BC, and documents from Mesopotamia indicate the use of cannabis as a drug at least by 500 BC. The ancient Indian civilisations of Mexico and South America used mind-altering chemicals such as cocaine, trilpines, harmines and indoles of various types. The natives of the Pacific Islands used beatel and kava kava, while in Asia natural preparations containing ephedrine and reserpine have been common for a thousand years. Closer to home we find a 200-year history of the use of opium; laughing gas or ether-sniffing parties, and cocaine epidemics during the last hundred years; and the consumption of alcohol has survived centuries of war and want, religious revival in Britain, and prohibition in the United States. So there is certainly nothing new about the idea of using chemical methods to alter one's state of mind.

Today thousands of chemicals are tested annually for their potential effect on the mind, or for their psychotropic properties as they are called. Expeditions have been launched as far afield as the Upper Congo and the Continental Shelf in search of new plants, animals or materials which might yield suitable chemicals. More psychotropic drugs come on to the market every year than any other type. It is difficult to name a single mental function unaffected in a definable way by some drug already available. What's more, we are getting better and better at making them. An opiate-like substance has been synthesised which is 10,000 times as powerful as morphine. Certain diazepoxides will produce sleep in doses as small as half a milligram. The legendary 'knock-out drop' may now be truly with us.

This spectacular development of the power and specificity of drugs stems from our increased understanding of the functions of the brain. Anti-depressant drugs which do not produce a false state of well-being, or euphoria, have been produced from tricyclic amines. They relieve depression by reducing the uptake of brain

Does Man have a basic need to intoxicate himself with drugs? Dr Tom Rusk thinks so. And that is why people smoke and drink.

norepinephrine in the storage granules of neurons. The faith held by pharmacologists that a person's mood and the chemical state of his neurons were one and the same thing seems to be at least well on the road to justification. But the development of a new drug is not possible unless people want it, and are potentially prepared to use it. Science alone is not responsible.

Anti-psychotic tranquillisers were introduced into the mental hospitals of the United States as recently as 1955. Since then the ever-growing number of mental patients in hospital has dwindled until it is now at the 1947 level. But most people do not realise that also in the United States aspirin is the second largest cause of acute drug death. Over the counter it is possible to obtain drugs that relieve tension, produce sleep, make you more alert, relieve all sorts of pain, reduce motion sickness, fight fatigue and so on. But the anti-histamines in common cold tablets can slow reflexes, and the 'safe, non-habit-forming' sedatives can induce severe hallucinations. Cases of caffeine poisoning from readily available tablets are by no means unknown. And finally there is alcohol. To call a drug a drink doesn't alter the fact.

At a London Conference of the British Medical Association called in July 1971 to discuss a voluntary ban on amphetamines Professor Sir Ronald Tunbridge, the Chairman of the Association's Board of Science and Education, said that an explosive epidemic of drug dependence had caught doctors unawares. 'Ten

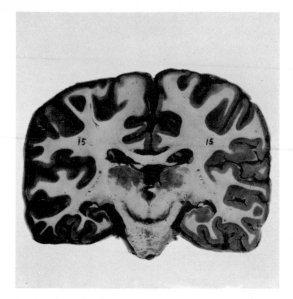
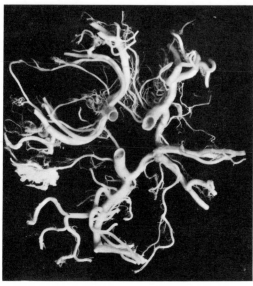

Section through a human brain.

Like tentacles, the major arteries spread through the brain to supply it with oxygenated blood.

years ago,' he said, 'we did not realise we would be facing this epidemic – an epidemic which is now of chronic dependence. We are now trying to make amends.' In fact, two years ago in the East Anglian town of Ipswich doctors adopted a voluntary ban on the prescription of amphetamines. Last year the BMA passed a resolution advocating a voluntary ban on their prescription. But so far only 39 of the BMA's 217 divisions have stopped prescribing them.

Sir George Godber, the Chief Medical Officer of the Department of Health, reports that at the end of last year the amount of amphetamines being prescribed in Great Britain has been reduced by one third. Even so this amounts to over 8 tons in England alone, mainly as slimming tablets.

Meanwhile in the United States in 1969, 90 million new prescriptions were issued for minor tranquillisers, 17 million new prescriptions for anti-depressive drugs, 12 million people had used marijuana at least once, and, to quote Wayne Evans of the American College of Neuro-Psycho Pharmacology, 'One calculates the consumption of diet pills, stimulants, aspirin, sleeping compounds and other psychotropic drugs by the box-car load. We have lived up to the famous comment – Man is the pill-taking animal.'

Amidst this welter of statistics, emotional reaction, and generally ill-disciplined thought and action, some straightforward sorting-out of fact from fiction is clearly essential if we are to get the problem into any realistic perspective. Dr Charles Tart, Associate Professor of Psychology in the University of California at Davis, is publishing a book called *On Being Stoned – a Psychological Study of Marijuana Intoxication*. It contains a scientific account of 37,000 experiences of marijuana smoking covering a total period of 450 years. It is the first document of its kind ever published. As Dr Tart points out, one of the difficulties of formulating objective opinions about self-induced drug intoxication is that the accounts tend to be anecdotal – they are stories heard at second or third hand.

Like many people Dr Tart predicts that the smoking of marijuana will be legalised in the United States, at least to some extent, within two or three years. He points out that no society is prepared to see the sons and daughters of its prominent citizens cast into jail for crimes the existence of which they do not recognise. When significant percentages of any peer groups in society set an example, it will inevitably be followed. Whether the veterans returning from Vietnam are regarded as heroes or not, they are certainly peer figures to young America.

In a beautiful garden on the coast of southern California a few months ago, a group of America's new, beautiful young people was assembled. The long hippy-style dresses of the girls were of genuine Eastern delicacy and splendour. Their leather was immensely expensive. The perfume which mingled with the hibiscus on the air was undeniably French. So was the champagne. The men were mostly lawyers. Their backgrounds were impeccable, their academic records of the highest. Their causes were the radical ones of the defence of Indian rights, pollution by industrial corporations, graft in high places. They were elegant, their conversation was charming and witty. Men and girls, they were all smoking marijuana, and they looked with distaste and surprise at a packet of English Virginia cigarettes. Said one of the most beautiful girls with a mocking smile, 'I am surprised at you, Mr Baxter. Those things give you lung cancer, you know.' This is in no way to dismiss or minimise the tragic results of drug abuse in the generally accepted sense, by an increasing number of young people throughout the Western world.

In the United States there is a growing body of scientific opinion that its spread should be regarded as the epidemic of a contagious disease, such as typhoid, and should be counter-attacked in the same way. One of the proponents of this view is Dr L. L. Judd of the University of California. Basing their studies on the way in which a contagious disease spreads he and his colleagues have been able positively to establish the relationship between the use of amphetamines and the follow-up of heroin. In areas where the use of methedrine by needle has become prevalent the use of heroin has developed very soon afterwards. They have been able to predict this sequence in specific localities and neighbourhoods. The remedial action has been to move in a team of specialists where the heavy use of amphetamines had been observed in the hope of preventing the subsequent heroin-addiction. The teams consist of ex-drug-users who have become professional therapists, psychiatrists who have become expert in dealing with drug-abusers, and a small support group.

The first requirement is to attend to the casualties that have occurred so far and to establish clinics on as informal a basis as possible. The target is to become involved with the drug-using population, particularly the leaders of it, and to win their aid in the treatment of others. They attempt to bridge the gap between the drug-abusing sub-culture and the established community, and then to attempt primary prevention via the school system. Although this method of operation is still in the experimental stages, results so far are encouraging. In San Diego county Dr Judd and his colleagues were able to establish a 'very good understanding of the drug-abuse pattern'. There was good community organisation and acceptance of their programme; they got to know who the people were, both users and dealers. What they attempt to establish is 'an anti-drug-abusing kind of culture'. Much of the operation is essentially 'without the law' but the objective is to get the drug-abusers themselves to close up the peddlers' business and thereby move them out of their community.

Dr Judd is convinced that decentralisation of the counteract against drug-abuse is of primary importance. He believes that the only fruitful approach towards prevention is via the drug sub-culture itself. He admits that his work is inhibited by current legislation: because the people with whom he deals are outside the law they have a very natural paranoia against anyone who is not a member of their own culture. Even drug-abusers from areas a short distance away are met with intense suspicion until they are proven to be acceptable. All this makes it very difficult to get at the root of the problem. Dr Judd and his colleagues would like to see the treatment of the drug-abuser removed entirely from the area of law enforcement and put within medical practice where it belongs. The victims are not essentially law-breaking individuals; they are obliged to break the law in order to support a habit from which they cannot escape. Yet, while most legislators agree that the treatment of drug abusers should be regarded as rehabilitation, there are those who are trying to make the law even more strict.

But at least the law in the United States is now beginning to discriminate (as it does in Britain) between the individual who is found in possession of enough drugs to meet his own requirements and those in possession of excessive quantities which would suggest that they were dealers. Dr Judd anticipates that within the next 2 or 3 years he and his colleagues could make a significant impact in San Diego county and possibly set up a model treatment-programme which might be followed by other parts of the United States. By then it should be established whether or not the idea of total decentralisation – of putting the treatment and training centres where they are needed, and not where it may be convenient to put them – is practical, and whether the training programme is effective.

Because of the experimental nature of the work care is being taken to achieve a scientific evaluation of the situation as it exists, and such progress as is made in combating it. It was found that in the 15 – 17 age group at High School 35 – 55% were using marijuana once or twice a month. In two College populations within the area, with an age group of 17 – 21, the percentage of marijuana smokers was not quite as high. In both groups the users of hard narcotics were about 3%. Approximately 1300 students were interviewed, so, although a figure of 1% addicted to heroin would be alarming, it is less significant in terms of the overall population.

Another important aspect of the research is the attempt to differentiate between the young person who experiments with drugs in a 'responsible way' and one who slips into the heavy dependence of the drug-abusing sub-culture. A study of 'drop outs', revealed that they come from all social classes, that their ages range from 16 to 27, averaging about 18, and that the majority were registered as students either at school or college. So most of them would have had access to medical care either through family physicians or through student health organisations, but that for indeterminate reasons they did not feel comfortable in presenting themselves for medical treatment under conventional circumstances. They would choose to travel many miles at considerable personal cost to visit a clinic more suitable to their particular requirements. But less than 5% of them gave as their reason either drug problems or psychiatric difficulties. The majority came for the treatment of venereal disease, for birth-control advice or appliances, or because they were pregnant. In essence, therefore, the clinic was providing a free medical service for middle-class youth for conditions which could be loosely described as 'discreet'.

In an attempt to explore ways in which they could be approached from the point of view of training and propaganda they were questioned on their attitude towards newspapers, television and the conventional, 'establishment' sources of information. Less than 3% of them believed anything that they heard or read coming from such sources. The implication is obvious: to make any impact on this extreme group, only the approach of the underground can be employed – the jungle telegraph upon which they themselves rely. It is this approach, therefore, which has to be followed if any attempt is to be made to come to grips with potentially epidemic areas. It is for this reason that in several laboratories in both the United States and Great Britain an analysing service is being made available so that drugs purchased on the street can be checked anonymously to ensure whether they are 'safe' or, because of adulteration or imperfection in manufacture, potentially lethal. To win their confidence, the drug-abusing sub-culture is being provided with its own highly scientific consumer-protection service.

The irony of this is in keeping with the recurrent paradoxes encountered in any attempt to assess our present knowledge of the brain and its function, and our attitudes towards this knowledge. Dr A. J. Mandell is a young research psychologist at the University of California with a personality as enthusiastically extrovert as Tom Rusk's. He was involved in a series of experiments in which day-old chicks were 'painlessly' blinded in one eye. (Should any reader be shocked by such an idea, he should ask himself when he last ate battery-reared chicken.)

The visual cortex was split so that the chicks had two independent optical systems, one for each eye, and a built-in control was provided thereby for experimental observation. It was anticipated that the chemicals associated with sight, the neuronal transmitters of the visual cortex, would decrease on the blind side, and could be objectively compared with amounts detected on the sighted side. The reverse was found to be the case. It was as if the neurons on the blind side were saying to their neighbours, 'Hey, where is the light? We must produce more enzyme to try to detect it.' This is precisely the opposite of the observable effect on muscles which decrease in size and atrophy in disuse, but expand and become more efficient if exercised.

Mandell suggests in conversation, although he is not yet prepared to put it in a scientific paper, that there is an apparent similarity to the lasting effect of massive and repeated doses of LSD on a human being. Like marijuana and the other hallucinogens, LSD is not addictive but repeated 'trips' do cause lasting personality changes. In the hippy commune there is no such thing as personal property. Furniture, sexual partners, even children are shared. They belong to everyone within the community, and to no one. In some ways this may seem an ideological situation but the integrity and significance of the ego is undermined. It is a state of affairs totally abnormal to human needs. 'What,' asks Mandell, 'can have happened to the minds of these people to make such a situation acceptable?' They have literally blown their minds: a permanent reverse mechanism has been established; the repeated super-stimulation and excitement of LSD or amphetamine makes them unnaturally placid, undemanding, emotionally 'dead'. Their minds fail to respond to the stimuli of normal needs and their gratification; they become not so much flower people as vegetables.

This suggests the apparent paradox that if you want the brain to do something chemically, you might ask it to do the opposite from that which is associated with the chemical you are looking for. Conversely, if you want to make a lasting behavioural change you make massive use of a drug which in the short term produces the opposite effect.

This 'behavioural engineering' is a frightening prospect. Mandell and many others foresee the establishment of boards of community leaders who might, for instance, be called upon to decide whether the aggressive side of an individual's character should be reduced in the public interest. That is not as far-fetched as may be supposed. Already the Courts in West Germany are empowered to order the administration of behaviour-changing drugs to habitual male sex criminals. At least this, to most people, seems preferable to the alternative, castration, which the German Courts can also order.

But the possibilities of behavioural engineering are not confined to the control of aggression and excessive sex drive, or the reverse. Another paradox which is the subject of heated debate, much of it misleading, is the likening of the brain to a computer, followed by attempts to compare their respective abilities. Computers' powers of calculation make the brain of the most brilliant human mathematician appear pathetically inadequate. They can grasp and manipulate thousands of facts in millionths of seconds whilst the brain's capability is measured in tens and tenths.

But if it ever becomes possible to express the moods of a normal sensitive human being in terms of a series of chemical formulae – and this seems likely – can we account for a musician's ability to memorise the complete score of several symphonies by a system of molecular biochemical coding in his head? A fairly simple computer could memorise every work in the entire musical repertoire, but as yet no computer has composed music which has the popular appeal of a Mozart Concerto, although compose they can.

Theorists suggest our memory code has developed its own symbolic short cuts. But others point out that even human speech is an inadequate means of communication, and that machines can communicate with each other millions of times faster and more explicitly than can humans. Perhaps by behavioural genetic engineering we could improve our human powers of communication, of creativity. But should we?

Robert Sinsheimer is Head of the Biology Division of Caltech and is an ardent admirer of A. A. Milne, whom he quotes at every possible opportunity. The opening lines of *Winnie the Pooh* may in fact have a new and profound significance for us all as we are brought face to face with this newly found knowledge about our own capabilities.

'Here is Edward Bear, coming downstairs now, bump, bump, bump, on the back of his head, behind Christopher Robin. It is, as far as he knows, the only way of coming downstairs, but sometimes he feels that there really is another way, if only he could stop bumping for a moment and think of it.'

Perhaps we may be on the verge of being able to win an escape from the head-bumping of the overwhelming problems with which we are now so totally pre-occupied.

It was the discovery of the effect of minute electrical currents in the brain which first began to lift the corner of the curtain over our understanding (see Chapter 8). This electrical activity, monitored by sensitive electrodes placed externally on the scalp and recorded by pen traces on paper – electro-encephalography or EEG as it is called, remains a key to our new-found knowledge. In fact its importance has been increased by two recent developments. The introduction of computer aids has increased the scope of the technique. And the new definitions of death, necessitated by mechanical and surgical ways to keep people's bodies working (heart-lung machines, cardiac manipulation, etc.), together with transplant surgery, have presented a social requirement for more specific scientific data on the meanings of life and death.

The 'brain wave' as recorded by the EEG is not the only measure of death of the brain. There are other equally important clinical indications, such as lack of spontaneous movement, lack of breathing, and so on. Nevertheless, the increasing importance being given to the EEG evidence of death became a cause for concern amongst neuro-scientists, including Dr Reginald Bickford, Professor of Neuro-sciences at the San Diego School of Medicine. As a pioneer and leading world authority in this field, Dr Bickford (who was born in Australia and was an early pupil and associate of the distinguished British neurologist, Dr Grey Walter) published in April 1971 a paper criticising certain aspects of EEG practice, particularly in the United States. In a subsequent paper he reported his research into the question of whether the EEG was really an adequate measure of death of the brain.

Doctors confronted by the pattern of traces from an EEG are now being called upon to look at a record of the brain wave, and to say whether or not the brain is dead. He considered that this was not a satisfactory approach, and has shown, amongst other things, that even an expert cannot make these judgements

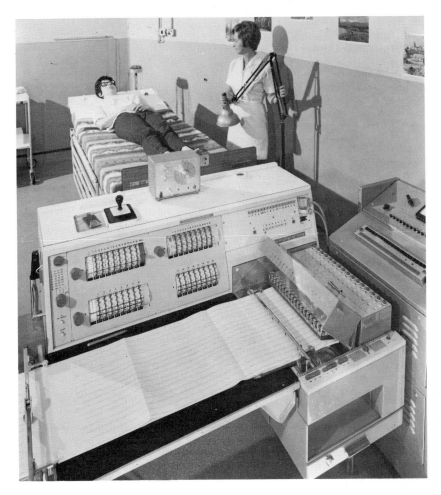

A modern 16-channel EEG machine for recording electrical activity in the brain. It is a most sophisticated machine capable of picking up the tiniest amount of activity.

adequately. Most records taken from a patient in whom the question of brain damage arises have a considerable number of heart waves in them because of the physical proximity of the head to the chest, and therefore to electrical waves produced by the heart. Dr Bickford's contention was that it was impossible to recognise the last remnant of brain wave when it was overridden by a heart wave. The trace was not 'flat' because of the interference of the heart wave, and therefore any pronouncement that there was no remaining evidence of an EEG trace could be nothing more than a guess. The heart could of course be kept going quite artificially.

In order to prove his point he put together in his laboratory electronic mixtures of brain waves and heart waves (EEG and EKG) of the kind that occur in brain-death patients. These were submitted for expert opinion without revealing the fact that they were artificially made laboratory studies. The experts were asked whether there were any brain waves in the record – in other words, if the 'patient' had suffered brain death. Fortunately for us all, it transpired that the experts were quite good, but even so in about 30% of these records errors were made in the interpretation. This was in traces when the EEG was low, but not so low that it wouldn't be critical in a diagnosis.

Dr Bickford's purpose was not to prove the inadequacy of doctors or electro-encephalographers called upon to make interpretations of this kind. It was to stress the inadequacy of the graphic presentation on which they had to make their

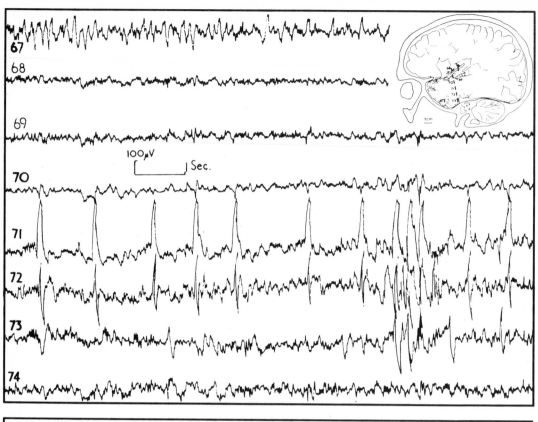

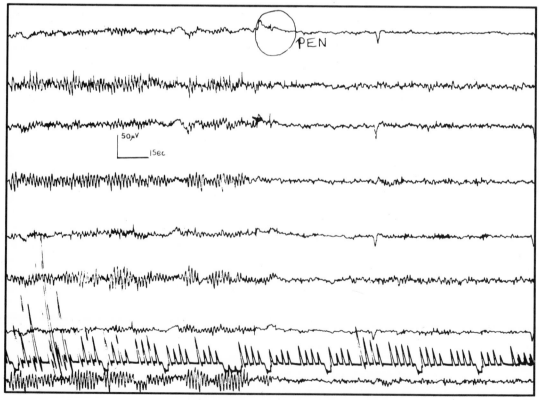

Electrical activity
recorded from the
brain of an epileptic
patient. The elec-
trodes that picked up
the signals had been
previously *inserted*
into the temporal
lobe during surgery.
The positions of these
electrodes are shown
in the inset. The high
activity in electrodes
71 and 72 indicates
abnormal brain tissue.

Electrical activity in
a normal brain. The
electrodes were placed
on the *surface* of the
scalp. The trace in the
first half of each line
indicates the alpha
rhythm which flattens
out and disappears
when the patient
opens his eyes.

vital judgements. The task which he and his team then set themselves was to improve the accuracy and legibility of the EEG trace in such cases, and to produce an EEG record from which the influence of the heart (EKG) had been subtracted.

In fact the entire process, although in outline it may appear simple, is an extremely complex computer task, the programming of which was developed by Dr Bickford and his colleagues. The computer stores various heart-beat patterns which could be described as a template. It uses about 200 before it can 'clean up' the brain waves. The result of this subtraction process could be written as a print-out in figures, but in fact what is produced is the computer's own 'cleaned-up' version of the EEG traces.

But Dr Bickford was still dissatisfied. The question which the electro-encephalographer is asked when called upon to define brain death is 'Does the EEG meet the criteria of brain death or not?' It is a yes/no question, and has nothing to do with how much brain wave there is in the record. It is simply, is there any? Bickford argued that a quantitive measurement should be introduced into this judgement. Instead of saying whether at any given moment it is one thing or the other, doctors should be able to watch the cumulative process – to observe the gradual diminution of the brain waves, eventually to the level at which there is nothing left. Again the computer was brought to bear on the problem. It observes and effectively measures the amount of activity revealed by the brain-wave traces and allocates to them an appropriate coded number related to its assessment of their quantity. The activity is expressed in microvolts, and this particular aspect of the case history of a dying patient might be that on the first day 30 microvolts of activity were recorded, 24 on the second, 20 on the third, 15 on the fourth, and so on until the reading was reduced to the level of 3 microvolts. However, any electronic system inevitably has a 'noise' level. It is a phenomenon which has bedevilled all kinds of electronic observation including advanced radar. The situation is like the effect of magnification in a microscope. There comes a point at which increased magnification does not improve resolution but has the reverse effect, and the image becomes blurred. Similarly, with electronic apparatus, when a search is being made for a minute signal, there comes a point when the increased amplification effectively buries the signal which is being sought.

It was found that the amplifier employed to seek out the last remaining trace of electrical activity in the brain could itself produce a trace which had the appearance of a brain wave. The irony of this is inescapable. The problem of defining the difference between life and death becomes a mathematical equation – the signal-to-noise ratio by a computer attached to an EEG machine and its amplifier.

In view of the complexity of the hardware, it became apparent that were this new technique to be put to practical use, some sort of time-sharing system for the computer would have to be evolved. But since the movement of patients in such marginal conditions between life and death is clearly undesirable, and the observations being made upon them are simply measurements of electrical activity, the next step was to establish a long-distance electrical connection between the EEG machine observing the patient's brain waves and the computer programmed to interpret its observations. Dr Bickford discovered a special tape recorder (developed for another purpose), which used tape so wide that it was capable of carrying 160 parallel tracks and had a head which could be switched to any particular track. On this tape was recorded a series of messages expressing the assessment of EEG values made by the computer throughout its range of capability. For example, 'Your patient has 30 microvolts of EEG activity'. This would indicate a fairly vigorous brain wave. Conversely, at the bottom end of the scale the prerecorded message would be, 'The EEG value at present is 3 microvolts. This is the noise level of the system and indicates that the EEG is isoelectric.' The patient is dead.

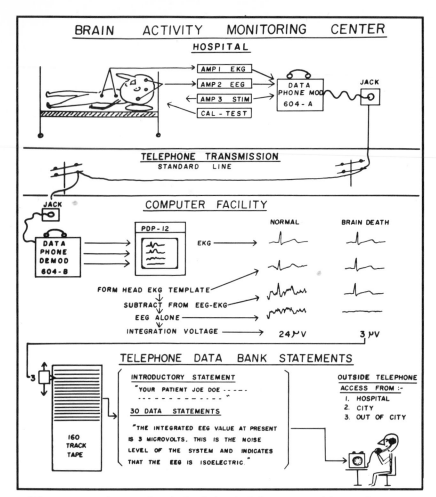

BRAIN ACTIVITY MONITORING CENTER

HOSPITAL

- AMP I EKG
- AMP 2 EEG
- AMP 3 STIM
- CAL - TEST

DATA PHONE MOD 604-A

JACK

TELEPHONE TRANSMISSION
STANDARD LINE

COMPUTER FACILITY

JACK

DATA PHONE DEMOD 604-B

PDP-12

EKG

NORMAL   BRAIN DEATH

FORM HEAD EKG TEMPLATE
SUBTRACT FROM EEG-EKG
EEG ALONE
INTEGRATION VOLTAGE   24 µV   3 µV

TELEPHONE DATA BANK STATEMENTS

160 TRACK TAPE

INTRODUCTORY STATEMENT
"YOUR PATIENT JOE DOE - - - - -
- - - - - - - - - - - - - - "

30 DATA STATEMENTS

"THE INTEGRATED EEG VALUE AT PRESENT
IS 3 MICROVOLTS. THIS IS THE NOISE
LEVEL OF THE SYSTEM AND INDICATES
THAT THE EEG IS ISOELECTRIC."

OUTSIDE TELEPHONE ACCESS FROM :-
1. HOSPITAL
2. CITY
3. OUT OF CITY

Dr Bickford drew this plan to illustrate his four-stage concept of a computerised brain monitoring centre. Brain waves from the patient are picked up by an EEG machine and fed by telephone line to a computer that has been programmed to analyse brain function. The reply comes back by telephone. A recorded computer voice reports the result.

The computer locates the playback-head of the tape machine to the message appropriate to the patient's condition and leaves it there until a further change in that condition is observed. The system is connected to the normal telephone service and the computer's 'voice' is thereby made available to anyone with access to the telephone number. The tape playback is the equivalent of a print-out, but as a simple enunciator it has the advantage of operating independently of the computer except when the need for an alteration of the message becomes apparent. In the interim the services of the expensive computer are available elsewhere. It is an uncanny experience to pick up a telephone, dial the appropriate number and hear a calm feminine voice say, 'The present level of EEG voltage recorded from your patient is two microvolts. This is within the lowest range, namely zero to three microvolts, which we recognise in patients. It corresponds to the noise level of the system and indicates an isoelectric EEG in which no brain activity is present. If this state persists for 24 hours and is not due to drug intoxication or hypothermia the EEG criteria of brain death or irreversible coma will have been fulfilled. End of message.'

The geographical location of participants in this train of communication could be almost world-wide. The computer and recorded messages are in the apparatus room adjacent to Dr Bickford's office just outside San Diego; the patient could be in Washington; the consultant surgeon receiving the messages could be in New York or London. Of course the greater the distance which the EEG signal is transmitted by telephone line from the patient to the interpreting computer, the greater the possibility of telephone noise interference. But it so happens that the frequencies in the brain waves of a dying patient are rather low, as might be expected, and in practice the noise problem has not proved to be too difficult.

Dr Bickford anticipates that similar enunciator systems could be developed for a wide range of applications for this kind of long-distance consultancy. Enzyme levels following a heart attack, for instance, provide a measure of damage to the heart. They could be measured on an automated system which could use a similar series of prerecorded statements to convey the information to any doctor by telephone. It is a cunningly simple method of providing a computer with a voice, without the highly complex electronics necessary for a truly 'speaking' computer.

Dr Bickford is now experimenting with even cheaper and more simple recording and playback apparatus. Having established that what he needs is thirty tracks for the messages of his EEG computer, he proposes to use eight-track commercial cassettes designed for stereo music, and by employing four cassettes have an availability of 32 tracks. His present computer is a $30,000 model, but, having systematised the computer's task as a result of his research, he is now confident that a more simple $5000 computer is capable of performing it, and he estimates the cost for a complete equipment at $10,000. He envisages the establishment of special centres for making brain diagnosis, which will not be confined to this relatively simple test of brain death.

Dr Bickford's questioning attitude towards the state of the art and science of electro-encephalography at the beginning of 1971 led him and his colleagues to devise a further new and important technique. Just as they use a computer to assist in the interpretation of EEG records and dissociate them from the peripheral effect of heart beats, they have also employed advanced computer techniques to improve and abbreviate the graphic presentation of the records themselves. The technique is called 'hidden line suppression', and it is a plotting trick which is programmed into the computer. The computer is instructed that no line which it has already drawn can be crossed by a successive line. If the tracing pen is about to cross an existing line, the computer makes it go round 'the obstacle' and resume its intended path on the opposite side of the intervening trace. The resultant effect is three-dimensional, as can be seen from the illustration.

An additional bonus, not part of the original intention, is the way in which the EEG information is compressed. In the past the clinician has been presented with a roll of paper perhaps running into a length of several yards, so confused and untidy in appearance that he is probably disinclined to read it anyway. Assisted by the computer, the same plotter can be made to condense its traces to 100 lines per inch without any loss of relevant information. In future doctors will be presented with one sheet of foolscap from which they will be able to see at a glance all the information required.

There are disadvantages to the technique because points of interest could effectively be hidden by the line suppression, but this can be avoided by rotating the display within the computer, so that the data can be viewed from different perspectives. The ultimate intention is to complete the automation of EEG by the computer including programme, interpretation and classification. But Dr Bickford is well aware that doctors may be reluctant to accept a computerised report unless they can check it by looking at some sort of trace for themselves. In the past, the clinician has had to rely upon the interpretation of the electro-encephalographer, but so concise and graphically simple are the presentations envisaged by the San Diego team that they predict the elimination of the EEGer in this role. As the heart surgeon likes to read his patient's EKG at the bedside, the brain surgeon will be able to do the same with the new-style EEG record.

Simplification, and therefore cheapness, are not the only advantages of this

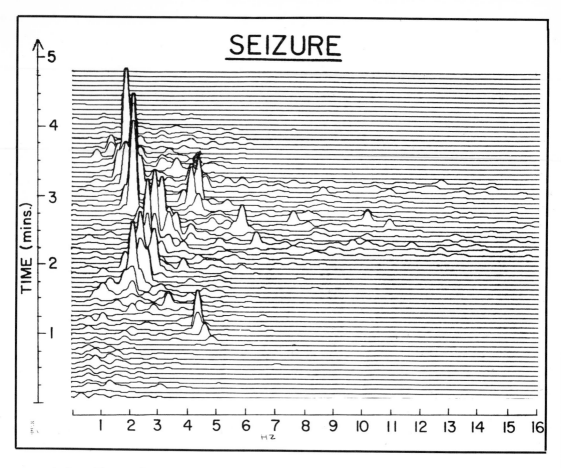

SEIZURE

TIME (mins.)

5

4

3

2

1

1 2 3 4 5 6 7 8 9 10 11 12 13 14 15 16

H z

new technique. Twenty-five years ago Grey Walter invented a frequency analyser and filter, based in an analogue computer which wrote on the original EEG record a spectrum rather like that produced by the hidden-line suppression technique. It proved to be far ahead of its time, and it only gave information on one channel, which limited its clinical usefulness. There was also insufficient knowledge available at the time to make use of the information which it presented, and the technique did not become popular. The new technique, however, presents a new depth of frequency analysis on every trace recorded, and although it is still in early infancy there is hope for its widespread adoption.

Dr Bickford's early association with Grey Walter in Britain is reflected in another aspect of his studies. It was Grey Walter who first demonstrated that a flashing light will create approximately the same rhythm of response in the brain. It's called photic driving. The response shows up very clearly in frequency analysis of EEG, and his study of it has led Bickford to suspect that the frequency response of an individual to photic driving may be highly characteristic and that, if presented in a three-dimensional way, it may be like a mental fingerprint. The traces produced by photic stimulation vary enormously from individual to individual. So far no one has been able to connect these differences with differences of personality, but they appear to be *unique* to each individual. If this can be satisfactorily established, it may become possible to identify people positively by their 'brain print' – a technique of considerable interest to 'spy-catchers' and less glamorous detectives mostly trying to identify a criminal who has deliberately changed his physical characteristics.

Computerised EEG. A dramatic representation of an epileptic fit (see page 219).

220

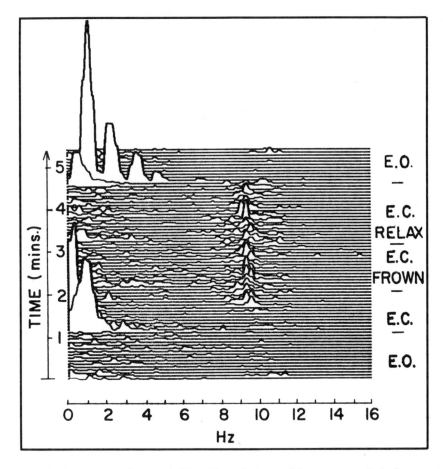

A further new study revealed by this technique of frequency analysis focuses on the alpha rhythms of the brain waves. It was the semi-mystical fascination to his students of the alpha rhythms which originally drew Dr Bickford's attention. In common with his colleagues he had often wondered why the human should have in his brain a rhythm as simple as the 10-cycle sound wave – the approximate EEG appearance of alpha waves. Subjected to the technique of frequency analysis, the alpha rhythms have revealed some very interesting and unexpected results. When the 10-cycle alpha rhythm was broken down into its quarter-, eighth- and finally sixteenth-cycle components, it was seen that these components behaved independently. What was assumed to be a regular response at 10-cycles proved to be the result of the addition of a number of highly complex traces. What had been more or less dismissed as something simple and uninteresting proved to be fantastically complicated, perhaps even containing some kind of language of its own. It may be that the analysis of the alpha rhythms will produce immense amounts of information, possibly even a 'brain information code'. The way to break this code would be to analyse, in the newly available range of frequencies, specific mental states such as meditation, concentration, visualisation and so on. If one component could be observed to change from state to state, the first step towards breaking the code would have been achieved. Already it has been established that everyone has a quite different-looking alpha rhythm, which suggests some relationship with personality. The possibility of associating specific thoughts with specific frequency responses is bound to stimulate renewed scientific interest.

Although the study of electrical activity in the brain has revealed a great deal

o

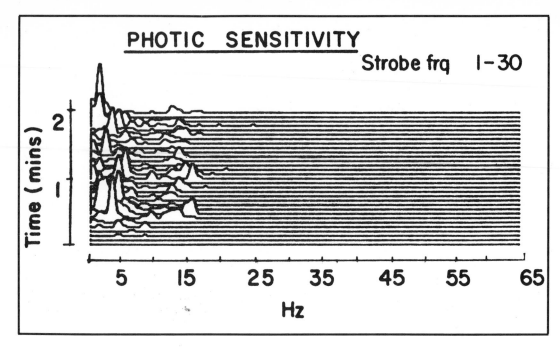

PHOTIC SENSITIVITY

Strobe frq 1-30

Time (mins) — vertical axis with markings 2, 1

Hz — horizontal axis with markings 5, 15, 25, 35, 45, 55, 65

of information about the function of specific areas and some mental processes, the long-held theory that the brain was a sort of electrically operated telephone exchange is now discounted. It has been established that the brain is not only an electrical device but also a chemical one. This resulted from the experimental work of S. P. Grossman and other researchers in the early 50s, who found that the introduction of minute quantities of chemicals dripped into the inner brain could cause remarkable changes in behaviour. Instead of applying the stimulation of electrical currents by electrode (see Chapter 8), Alan Fisher dripped male hormone into the brain of a male rat. He expected a response in terms of sexual behaviour, but in fact the male rat began to exhibit pronounced female tendencies, tried to mother a female rat, and set about building a nest. When the hormone was applied to a slightly different point in the animal's brain, the expected stimulation of masculine sexual behaviour resulted. Similarly it was discovered that females could be induced to act like males. Researchers went on to demonstrate that different substances dripped into the same place in an animal's brain would produce different effects. That specific groups of brain cells or neurones are affected in particular ways by particular chemicals suggests that the brain function is a matter of chemistry. Enzymes, the substances which regulate specific biochemical processes, are found to be distributed in distinctive patterns throughout the brain, and this suggests different chemical activity from area to area. Since the function of specific areas had been established, the connection between specific chemical processes and the appropriate function became apparent.

The discovery by James Olds in 1953 of the pleasure centres of the brain by what could be described as electrical exploration had startling philosophical implications. The definition of difference between the physical brain and the inanimate Mind of Man has been the subject of long and embittered debate. It is rapidly becoming increasingly difficult to make any differentiation whatsoever. Some of the most recent biochemical research into brain function now suggests that memory – the key quality of any 'mindful' activity – may be nothing more nor less than protein.

Mental fingerprints. When you look at a flashing light your brain reacts in a manner unique to you alone.

222

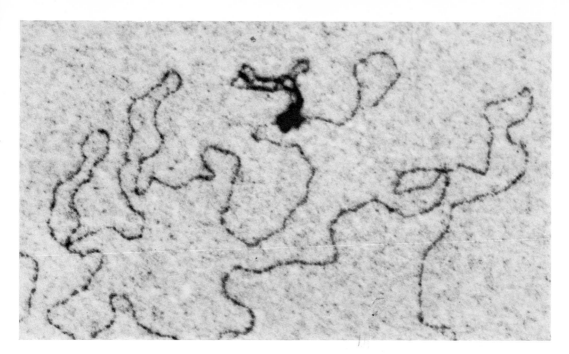

Strand of DNA magnified nearly a quarter of a million times. The black area at the top of the picture is a concentration of RNA with enzyme activity going on. This extraordinary photograph was taken on an A.E.I. electron microscope at the Basel Institute for Immunology and it has never been published before.

The fact that genetic information is transmitted as a code in the molecules of DNA and RNA makes plausible the belief that other types of information might also be coded in similar proteins. As Dr Bickford suggests the possibility of coded information hidden within the electrical alpha rhythms, so microbiologists are now seeking the biochemical molecular code of those neurons, or brain cells, associated with the function of memory. In this type of research the brains of trained animals such as mice and rats are studied. The training to which they are subjected is extremely simple – the association of the performance of elementary tasks such as the pressing of a lever with reward (food) or punishment (a mild electric shock).

As a by-product of their research a team of researchers in California have produced a 'mouse-training machine'. Five or ten mice at a time are put on to a revolving platform, each in his own small compartment, each confronted by a choice of lever to push. The platform rotates in a random fashion, and the response of each animal is recorded automatically by computer. The process is continued until a satisfactory performance is recorded – a matter of a few minutes only – and another batch of trained animals has been automatically prepared. The 'training' is necessary for the study of their memory, since the recorded pattern of behaviour can be compared with their performance in subsequent experiments. Chemicals associated with the brain are injected into the tails of the mice before, during or after the training process, and the effect upon the animal's memory is demonstrated when the mice are called upon to perform the same task subsequently.

One of many scientists engaged in these studies is Dr Samuel Barondes of the University of California. At the age of 37 he already has a brilliant career in psychiatry behind him. The belief is that the way in which brain cells regulate each other is by making proteins. The protein synthesis required for the psychological function Dr Barondes is studying – memory storage – 'seems to be going on in neuro-cells and to be an expression of the mechanism by which these cells regulate each other'. The protein is developed within the cell and is made at a greater rate in a stimulated cell than in a non-stimulated one. The process could conceivably be observable on an electron microscope, but as yet it is impossible

to see protein molecules that are above 'background'. In fact it is possible under unusual circumstances to discern isolated protein molecules, but the problem in observing change as a result of behaviour is that, as the cells are made up of protein anyway, you wouldn't expect to see anything different.

DNA the complex helix that makes us what we are.

Memory is clearly related to behaviour if behaviour is the product of experience, since experience is inextricably associated with memory. Barondes agrees that the type of memory inhibitions which he can produce must therefore have their personality effects of one kind or another. But he does not agree that this could be described as 'personality engineering', in the same way that other aspects of current research have been described as 'genetic engineering', for, if one trains an animal to do something, one has modified his behaviour irrespective of what is going on biochemically in his head. He believes that with the chemicals that he and his colleagues are using, they could engage in personality engineering in the sense of blocking the memory storage processes. However, the real purpose of this work is to discover more about specific brain proteins.

They may provide the key to our understanding of the relationship between neurons. 'How does one neuron recognise another and say, "Yes, I am going to make a connection with you", and to another, "No, I am not"?' asks Dr Barondes. He and his colleagues are currently conducting a series of experiments with glycoproteins, which are abundant on the surface of brain cells. It is known that recognition functions are achieved on cell surfaces, and in view of their abundance it therefore seems likely that they play some role in cellular recognition.

Inter-neuronal relationships, the ways in which one brain cell communicates with its neighbour, explain how the whole brain system is 'wired up'. This is not only the subject of major scientific curiosity, but also controversy. The now old-fashioned idea that it is purely an electrical process still has its adherents. In *Encounter* in February 1971 Gordon Rattray Taylor adopts a writer's compromise. Referring to the long-established argument as being between the 'sparks and the soup men', he affirms that they were both right. 'An electrical impulse travels along each nerve fibre; at the end of it, it releases a flood of chemicals [transmitter substances] which cross the gap to the next neuron, either stimulating it or inhibiting it. Thus the chemical and electrical behaviour of the brain are closely integrated.'

In *The American Scientist* a month later Dr Patrick McGeer, head of the Laboratory of Neurologic Research in the University of British Columbia, writes: 'It has now been established beyond reasonable doubt that communication between the neurons . . . is by means of chemical agents, or neuro-transmitters, which are released from the terminal of one neuron and cross the synaptic cleft [the space between the two cells] to influence the excitability of the next neuron. The alternative possibility of electrical transmission has been ruled out.' McGeer suggests that the existence of these chemical messengers provides a possible way of influencing behaviours and mental performance while leaving other aspects of brain function almost completely unaffected. If the transmitters governing the cells associated with such functions as sex, appetite, sleep or mood turn out to be specific, and if chemical methods can be found for selectively interfering with their activity, then fairly precise behaviour modifications might be brought about. While not denying this possibility, Barondes and his colleagues have yet to achieve such specific results from their research. They have proved that they can inhibit the memory of animals, but not in a specific way.

The first chemical approach to the problem of memory was made in 1959 by Hyden, the Swedish scientist who endeavoured to detect a molecular record of acquired information by measuring chemical changes in brain cells involved in learning. Some of his observations were confirmed elsewhere by scientists who observed an increased rate in the production of RNA and protein in the brains of animals during learning. Subsequent investigations have demonstrated that these chemical changes are necessary for the retention of such acquired information (i.e. memory), since learning can be impaired by the administration of drugs that inhibit the production of RNA or protein.

Although these experimental observations suggest that learning is accompanied by chemical changes in the brain, they do not prove that these changes represent a record of memory. Here again we find current scientific controversy. The majority view is that the chemical changes observed are non-specific, i.e. the substances produced by the cells are the same, irrespective of the precise nature of the information required. They are probably enzymes necessary to the transmitter-substance that bridge the cell connections involved in the learning process.

The alternative theory, supported amongst others by George Ungar of the Baylor College of Medicine, is that the increased protein-production observed during learning includes the making of coded molecules which represent a record of the newly acquired information. This process could be directly compared to the molecular code of genetic information now positively associated with DNA. The molecular coded memory, however, must be based on a different system of signals. Memory is a record of individual experience. We neither inherit it nor transmit it to our descendants. One of the most important studies of the Austrian scientist Konrad Lorenz at his Parkland nature conservatory near Munich has been to delineate the difference between acquired (learned) and innate (inherited) behaviour in animals, particularly wild-fowl.

The brain function involves a continuous traffic of incoming stimuli and outgoing responses. Trying to identify specific chemicals related to the acquisition of specific information, against this background of chemical 'noise', is therefore even more of an overwhelming task than Dr Bickford's problem in recording vestigial brain waves. No chemical methods known today are sufficiently sensitive and selective to detect the unique substance representing a given memory trace. This explains why the results of chemical analysis of the brain have so far remained inconclusive.

In 1962 at the University of Michigan psychologist James McConnell conducted a series of bizarre experiments with a certain type of flat worm. He taught them to curl up in anticipation of an electric shock when a light was flashed. He then ground them up and fed them to untrained worms. The worms that had eaten the trained ones learnt to contract twice as quickly as their predecessors.

Georges Ungar undertook a similar series of experiments. He conditioned rats to shun the darkness which they normally prefer, and then injected an extract from their brains into mice. These subsequently demonstrated a marked preference for light in contradiction to their 'natural' instinct. The work was followed up in laboratories in Czechoslovakia, in Denmark and the United States, and by 1970 similar experiments by at least 28 laboratories had confirmed the results. A wide range of learnt behavioural patterns have been successfully transferred in this way, involving the conditioned avoidance of light and noise, preferences of choice for degrees of brightness or space, the response to rewards, and other comparatively complex behavioural characteristics.

In a brilliant feat of chemical detective work Georges Ungar then succeeded in isolating a minute quantity of the substance in the rat-brain extract which was actually responsible for the fear of the dark engendered in his mice, which he called scotophobin. In the course of this analysis his disproved the theory that the active material was RNA, and demonstrated that the activity of scotophobin was not affected by treatment with ribonuclease, the enzyme which breaks RNA down into its constituent nucleotides.

Ungar believes that scotophobin is the first to be isolated of a very large number of biochemical substances into which experience is converted for subsequent processing and storage in the nervous system. As a result Dr W. Parr of the University of Houston has made a synthetic compound the activity of which is identical with that of the natural scotophobin isolated from animal-brain extract.

Any code is a system of signals representing information. Language, writing, tape recordings and maps are commonplace examples. When the rules of the code are known, they can be reconverted into the original information. It therefore follows that when the rules of the neuro-code are known, the information encoded in the active biochemical substance could be read out. It is even possible that we could synthesise molecules by which any new type of information could be introduced into the brain and cause the corresponding behaviour. But this is looking far into the future of neuro-science. The number of code substances probably runs into billions, but it would be unnecessary to know all these code 'words' in order to break the code – only the rules by which they are assembled.

226

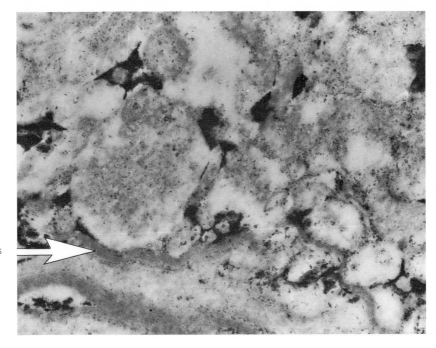

Enzymes at work in the nerve endings within a brain. The enzymes are the black areas.

Nerve endings

It can be assumed that the alphabet of the code consists of the twenty amino acids that make up all the proteins present in animals. What is already known about the brain during embryonic development provides some indication of the basic vocabulary. During prenatal growth the neurons are assembled by a mechanism of chemical recognition to form the different pathways of communication within the brain. Neurons bearing the same molecular label organise themselves next to each other and create a pathway through which specific information is transmitted. They are fully organised at birth, but some of them require further development before they become fully functional.

William Schumaker, a graduate student, and Dr Richard Wurtman, the professor of Endocrinology and Metabolism, at MIT's Department of Nutrition and Food Science published in March of this year direct biochemical evidence of damage to brain neurons due to malnutrition during the few weeks before and after birth. They found that the brains of rats that had been under-nourished from mid-gestation and were killed at weaning contained 25% less norepinephrine than the brains of adequately fed control animals. Norepinephrine is a neuro-transmitter, i.e. it is active in the connection between adjacent brain cells.

Dr Wurtman calls the findings a first small step in understanding how protein malnutrition affects brain neurons. There has been mounting evidence that inadequate protein in the early life of animals and people interferes with the development of the brain – with the ability to learn – and that it affects behaviour. Several scientists have discovered chemical changes in the brains of malnourished animals and children, but these changes were concerned with the amounts of DNA and other substances present. The exciting point in the MIT studies is that a decrease in brain norepinephrine can only result from the activity of the neurons themselves. Neurons which release norepinephrine are known to have a role in the control of mood, in learning, the regulation of blood pressure, heart-rate, blood sugar, and glandular function. Another neuro-transmitter, dopamine, was also found to be deficient. Low dopamine levels in the brain have already been positively associated with Parkinson's disease.

The work of Schumaker and Wurtman is likely to be of major importance in the study of human diseases. Kwashiorkor, for example, is a condition of acute protein starvation which is found in children between the ages of one and three in underdeveloped areas. Even when they grow up such children suffer from behavioural defects and impaired learning ability. Preliminary results of research on kwashiorkor in Guatemala suggest that under conditions of in-adequate protein early in life there is a low level of norepinephrine in the human brain comparable to that found in the experimental rats.

Dr Wurtman has not yet established if the low level is because there are fewer neurons containing the substance or because there is less norepinephrine spread amongst a normal number of neurons. If the number of neurons is normal, the deficiency might be reversible by correcting the dietary protein or by drugs. However, if the neurons are low, the condition might be incurable, since brain neurons lose the ability to divide long before the rest of the body is mature. Future studies with rats and other animals may provide an insight into the reversibility of these norepinephrine deficiencies.

It is only during the past 20 years that the presence in the brain of these bio-genic amines, as they are called, has been discovered. In the past 5 years the catalogue of known neuro-transmitters or chemical messengers of the brain has been built up to include dopamine, noradrenaline, serotonin, and acetylcholine. These are highly localised to those areas of the brain which physiologists had long associated with mood, sex, appetite and primitive movement functions. Their presence has made it possible to map the pathways in the brain associated with them and to classify those pathways biochemically and functionally.

The pathway that has received most attention to date is that containing dopa-mine. It extends from the Substantia Nigra (black material) to the Caudate and Putamen. It is this system which is intimately involved with extrapyramidal or involuntary movement, and which appears to be defective in Parkinson's disease. By a brilliant combination of observation, analysis and applied logic – the classic method by which all true scientific advance is made – it was established that the message-carrying function of this pathway was at fault. The low level of dopamine was ascribed to the inability of the neurons to manufacture it in sufficient quantity, although the reason for this remains obscure. Meanwhile a method for compensating for the deficiency has been devised: this is to administer L-dopa, the precursor of dopamine. Unlike dopamine itself, dopa can cross the blood-brain barrier readily and was at first tried in small doses in cases of Parkinson's disease in 1961. Larger doses were tried by other researchers in 1964 but the results were not particularly striking, and it was concluded that so many cells had been permanently damaged in the condition that the inactive dopa could not be converted in sufficient quantities into active dopamine. But in 1967 even larger doses of the more active L-dopa were administered and spectacular results were immediately achieved in patients who had been almost totally incapacitated for years by this dreadful disease in which paralysis and involuntary twitching and shaking are amongst the more distressing symptoms. L-dopa is now re-garded as 'the agent of choice' in treating Parkinson's disease.

Quite apart from this triumph over a terrible disease which had hitherto been regarded as incurable, the treatment has important implications in the theory which lies behind it. It is difficult to imagine that such results could have been achieved if deficient cells had been irreversibly damaged; it is more likely that many are merely under-performing owing to an inability to manufacture sufficient dopamine. In other words, the severe and disabling symptoms of Parkinson's disease may not result entirely from irredeemable cell loss but may partly be due to functional impairment owing to a lack of an essential enzyme.

The logic and precision of biochemical attacks upon illnesses of the brain such as these provide a marked contrast to the inevitably destructive techniques of surgery which had until recently been regarded as the only possible approach. Nevertheless it must be admitted that, without the comparative butchery of the surgeon in both clinical and experimental techniques and the observation of the results of appalling brain injury resulting from accidents and war, we should not have reached our present state of knowledge about the functional geography of the brain. In the 1940s, particularly in America, a large number of operations were performed on mental patients in which large sections of the frontal lobes of their brains were either removed or rendered inoperative. Although the undesirable characteristics which resulted from the illness of these unfortunate people were removed by the surgery, many were left in an equally pathetic condition. In one extraordinary and thoroughly documented case the memory of a young man was totally destroyed. He is apparently normal in all respects but for the fact that he lives from minute to minute with no recall whatsoever of the past. He will read over and over again the same article in a magazine with no loss of interest. For him each reading is his first. And yet recently, and with no apparent effort, he recognised the head of President Kennedy on a coin, and stated that the man had been assassinated. This may be an example of the extraordinary way in which the function of one sector of the brain, which may be damaged, can apparently be assumed by other areas. A well-known example of this in Britain is the case of Stirling Moss the former racing driver. The brain damage which he sustained in his last accident at Goodwood was extremely extensive. For a while, to those who knew him, the effect of this damage was apparent. Today it would be impossible for anyone in the presence of this extremely active personality to observe any possible residual effect of his injuries.

In Russia the case of a former Officer of the Red army, Lev Zassetsky, is far more remarkable. During the Second World War part of his head was shot away. His personality was not changed but his vision was affected and he lost his ability to read and write. His memory was also impaired but his courage never deserted him. Under the care of Alexander Luria of Moscow, one of the world's leading specialists on the mental effects of brain damage, Zassetsky has learned to read and write again, presumably using unorthodox brain mechanisms for the purpose, and has similarly compensated in many ways for the impairment to his capabilities. Over 25 years since sustaining his injuries he has now written a manuscript of 3000 pages of autobiography.

Apart from the personal courage and tremendous efforts of will in such cases as those of Moss and Zassetsky, there is almost certainly a biochemical explanation for their recoveries. And in contrast to the surgical approach to mental illness (as opposed to accidental or war brain damage) the biochemists are now focusing their attention upon the chemical causes and consequent chemical treatment of mental disturbance such as schizophrenia. In January 1971 in Philadelphia Dr Larry Stein and Dr David Wise published a report which may lead to the development of a completely new approach to the treatment of mental illness. This report could fairly be described as biochemical psychiatry. As long ago as 1884 W. A. Thudichun wrote in London: 'Many forms of insanity are caused by poisons fermented within the body.' Echoes of Dr Joseph Issel's theories about cancer are inescapable. But in this case the essential properties of the offending chemical were scientifically outlined by Professor L. E. Hollister in 1968 as 'a toxin highly active and highly specific in its action at minute doses, continuously produced, for which tolerance does not develop'. Now Stein and Wise believe they may have named the culprit – 6-hydroxydopamine – a molecular variant of the dopamine already referred to.

Schizophrenia is an appalling, often life-long illness, in which the sufferer appears to shut out the world about him totally, paying no attention whatsoever to anything for years, not moving, not speaking, apparently not seeing. Their thoughts do not seem related and directed by any purpose or towards any goal. Although the effect of the drugs LSD or mescaline have been compared to schizophrenia, Stein and Wise point out that the wide variety of mental changes caused by the drugs, such as hallucinations and delusions, tend only to resemble the accessory symptoms of schizophrenia and not the basic characteristics of the disease.

The early impression that schizophrenia is inherited has been proved. Systematic family studies have established the importance of genetic factors in the development of the disease 'beyond reasonable doubt'. This necessarily implies that it must be caused by a biochemical malfunction since no other mechanism is known for the reproduction of genetic tendencies. Observation of schizophrenics, and experiments involving the electrical stimulation of the brain, suggested that the area at fault was the pleasure centres (see Chapter 8). Rats given the ability to apply electrical self-stimulation to their pleasure centres were injected with 6-hydroxydopamine. They immediately lost interest in the pleasure-giving lever which they had previously pressed so energetically. This suggested that the chemical had blocked the effect of the pleasure-sensing mechanism. But when the rats were protected against the effect of this substance by a pre-injection of pure chlorpromazine hydrochloride, the effect of the hydroxydopamine was nullified and the rats continued their pleasure-giving activity.

It was already known that 6-hydroxydopamine produced in the body can induce the degeneration of nerve terminals and cause a marked and long-lasting depletion of norepinephrine. Schizophrenics were known to have exceptionally low levels of this substance in their brains. The theory therefore evolves as follows. As a result of an inherited enzyme defect, 6-hydroxydopamine is occasionally or continuously made in the body. This causes a biological disturbance which limits the amount of norepinephrine in the brain which is essential to the function of the pleasure areas. But the introduction of chlorpromazine may act as a sort of antidote and prevent the damage which would be caused by the 'toxic' substance.

Chlorpromazine had already been proved a useful drug in the treatment of schizophrenia although the reason for its desirable effect was not fully understood. It would now appear that it protects the reward system of the schizophrenic's brain by blocking the uptake of the 6-hydroxydopamine formed in his body. Although the formation of 6-hydroxydopamine may continue in a patient so treated, the toxic substance no longer has entry to the vulnerable site. This theory also accounts for the fact that chlorpromazine as a treatment drug is less effective in the case of 'burnt-out' schizophrenics. The assumption is that in these cases the norepinephrine reward terminals have already suffered irreversible damage. A further thread of evidence in support of the suggestion that schizophrenia is caused by 6-hydroxydopamine is the isolation and identification of a substance with a highly characteristic smell in the sweat of schizophrenic patients.

The ability of certain chemicals to 'block' neuronal pathways is used extensively in research and in the treatment of disease. The blocking process itself is also a subject of major study. A. H. Black, Professor of Psychology in McMaster University in Canada, has used chemically-induced blocking and electrical stimulation of the pleasure areas to induce specific brain activity, as observed by EEG in experimental animals. He has 'trained' them to produce specific brain-wave patterns on command. It is true that if in the normal way you train a dog to 'sit' he will presumably produce the 'sit' pattern of brain waves observable by EEG in order to win his reward of a biscuit or pat – or a scolding or smack if he persistently refuses to obey. You could induce a child to practise the piano in the same way. But the resultant brain waves would not be sufficiently specific for scientific study. All kinds of other responses would be involved, including emotional ones – the desire to please. This would be in conflict with the desire to do something else – run about, or go out to play with friends. There would also be the reaction of pleasurable anticipation of the reward as well as fear of the possible punishment. This 'anticipation response' is one of the more baffling objects of a separate and very complex study. Cats, for instance, have been observed not to bother to 'collect' their reward stimulus, apparently deriving sufficient pleasure from its anticipation.

However, in this particular series of experiments Black 'blocked' out these other factors chemically, and succeeded in training animals to produce specific types of neural activity on command by what is known as operant conditioning – the *immediate* presentation of reward (by pleasure stimulation) following the production of the desired mental response. To the average person that may seem an extraordinarily complicated way of demonstrating the obvious, i.e. that the brain can be trained to think specific thoughts. But what Black is getting at is the vast area of brain acitivity which has been considered unapproachable – the automatic and subconscious areas which control heart-beat, blood pressure, breathing and so on.

Black admits that he is still a long way from a satisfactory understanding of the theta waves of electrical activity from that section of the brain called the hippo-campus – the object of his particular study. However, since this activity does seem to be related to the neural processes that control certain classes of behaviour, it would seem that certain important psychological and neural processes are being changed when certain patterns of electrical activity in the central nervous system are produced.

An American researcher, J. Kamiya, reported to a Californian conference in 1969 that using human subjects his experiments suggested that he could alter mood states by rewarding appropriate brain-wave patterns. People conditioned to produce alpha rhythms, the subject of Dr Bickford's studies reported at the beginning of this chapter, said that they were in a 'tranquil, calm and alert state'. Kamiya speculated that this might be similar to Zen and Yoga meditation and suggested that operant conditioning may be a more efficient way of learning to achieve this state than the usual protracted training procedures – a method perhaps for attaining instant Nirvana – a highly developed meditative state considered one of the most desirable conditions achievable in these cults.

The suggestion is that new and more efficient types of control can be achieved by rewarding brain waves. On the other hand it has yet to be proved that such conditions cannot be achieved just as effectively by rewarding the readily observable outward condition which they would produce. It also still has to be established to what extent the electrical pattern of neuro-activity can be influenced. It is of course incorrect to assume that we can condition *all* patterns just because we can condition *some* of them. The overall emotional state of the person under observation is also relevant: we might be able to condition one pattern only when he is serene and calm, and another pattern during a much wider variety of emotional conditions. Indeed, can conditioning be achieved outside the labora-tory? There would be little point in curing a stutterer in the consulting room if as soon as he left his speech defect returned.

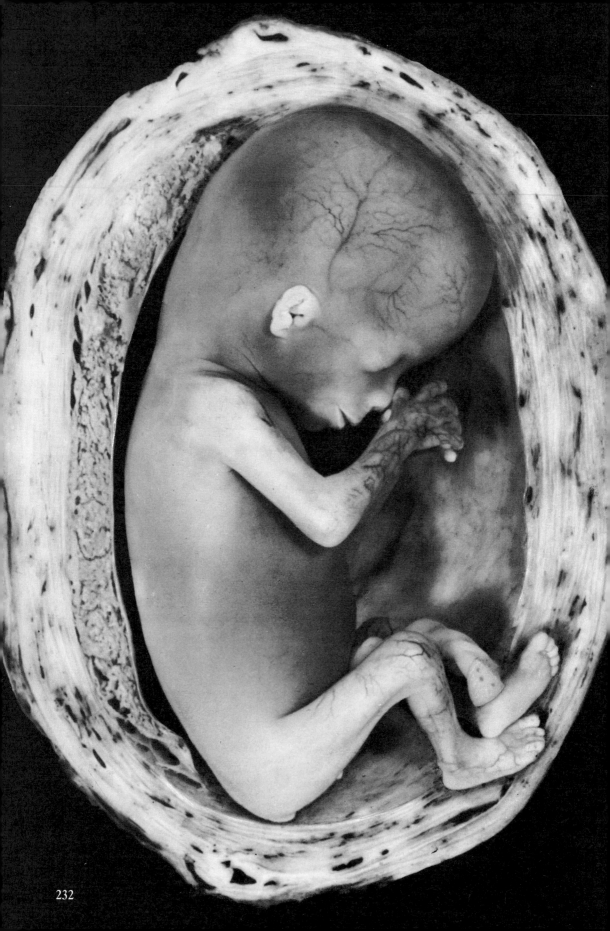

10   In Sickness and in Health

The most disturbing thing about both science and technology is the apparently haphazard way they develop. Research inches forward along thousands of different paths. As the work specialises and increases in complexity, communication from one sector of research to another becomes difficult, if not impossible. Today there can be hundreds of scientists beavering away on the same group of problems, and in many cases most of them do not even know that the others exist. There is no overall control, or even cataloguing of activity. So time is wasted on many ideas discarded because they do not happen to fit the particular pattern of the individual researcher's thoughts. How often, then, might there have been a great discovery, a whole new science, in the pile of theories and tests rejected because apparently they didn't lead to what one man thought the goal should be? Equally, how often has accident, or a television programme, brought the work of two scientists together, for a meeting that produced answers both of them might have spent fruitless years looking for in the wrong way? That example of cross-communication is rare. As specialist fields become more refined, and multiply, it becomes rarer.

It is this increase in specialisation, and the impossibility of achieving an overall view of science and technology, that leads to the greatest paradox: one end of the laboratory, so to speak, working against the other. As everybody forges ahead in their own tiny worlds, blinkered by the very need for specialisation that makes their work viable, scientists create problems for each other. In one way, they create those problems almost deliberately: today in America, where laboratories are involved in a fight for dwindling financial support from government or private industry, it is often not in a researcher's interest to let his colleagues know too much about his work. To do so may help a rival capture a precious grant. There is an increasing suspicion among the general public that, far from ushering in the Utopia that they were once thought capable of creating, science and technology have become so involved in development for its own, often commercial, sake that they are leading us towards a kind of anarchy. One eminent American has suggested that it is already here. He sees an increasing and bewildering isolation for the man in the street, who, with his own very simple needs and desires, is being force-fed with new developments, new ethical problems created by the researchers that he has neither the ability nor the will to comprehend. In engineering establishments, drug houses, chemical and physical laboratories everywhere, no sooner do researchers bring out a product or a development that will affect all our lives than they are at work on another. We are, in a sense, taking, using and passing on to our descendants articles of social change that we ourselves only dimly comprehend. But our great-grand-children may look back at the last three decades of the 20th century and blame medicine for being the greatest source of the evils they will have to live with. In its haste to use the glittering new tools of science to further the aim of a 2000-year-old law that a doctor must preserve the life of an individual, medicine has undoubtedly led us towards nothing less than a total reassessment of the value of life itself. Sadly, there is no indication that we are even capable of handling the kind of decisions that medical science is about to put into our hands. Yet at the same time there seems to be no way to put off those decisions. We are presented with a *fait accompli*, because in a sense there can be no going back. How, in the

By the end of the third month the foetus – now technically no longer an embryo – is 70–80 mm long. It has doubled its length in the previous four weeks. Now it is large enough for individual characteristics to be distinguished with the naked eye (see page 235).

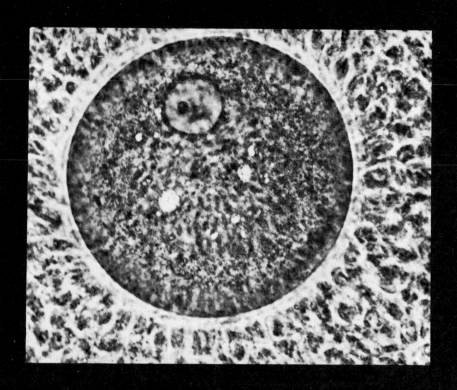

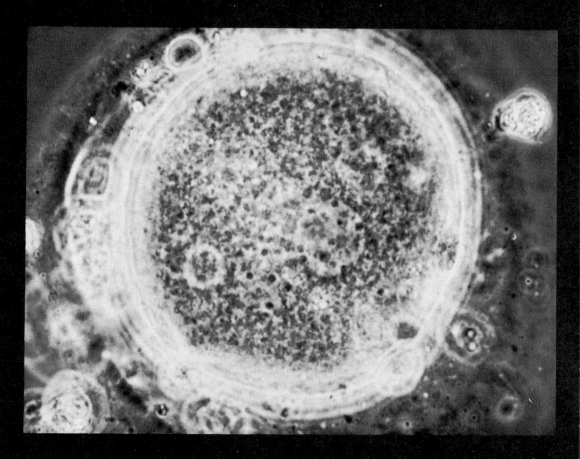

welter of medical experimentation going on all over the world, would we begin to say 'discontinue this work, concentrate on a different aspect of that research'? Each experiment has its own reason for being, each one is as valid in its own context as the next. So a *fait accompli* cannot be presented by some central authority that can preventively disestablish or destroy. It will arise from the accidental coming-together of many different lines of research, and we will not know the effect the event will have on our lives until it happens. At that point it will be too late. The best we can do is look at some of the proudest achievements of medical science today, and prepare ourselves as best we can.

One biological technique which may have the most profound effects before the end of the century exploded into the headlines a few years ago when a group of researchers succeeded in taking the eggs from a woman, keeping them alive for a number of hours in a laboratory environment, and then fertilising them with live male spermatozoa. Nobody has since succeeded in taking this development to its logical conclusion, to produce the 'test-tube babies' that were predicted on every front page at the beginning of the sixties, when the idea first began to look feasible. What *has* happened, however, is that researchers have found a different test-tube – another woman. The reason for taking the egg out in the first place was to develop ways in which a woman might be able to 'foster' her eggs within another. The system developed, and now nearing perfection, is called 'egg transfer'. It has been carried out on many occasions with animals, as far back as 1890. If the woman is capable of having her eggs fertilised, but not of carrying the foetus to full term, then in a sense the problem is relatively simple. The fertilised eggs are placed in a foster womb, and the baby grows to term, and is born to the carrier. Though that carrier has fed and nurtured the foetus for 9 months, in every other sense the baby is the son or daughter of the original mother and her husband. Little, however, is known about what happens to the baby in the womb, in what way the borrowed eggs might be affected by developing within a foreign, though living, organism. What will the foster-mother 'give' the visiting foetus while it is inside her?

The ethical problem increases in complexity when the technique becomes available to women who are barren. In this case borrowed eggs could be taken from another woman, a fertile donor. It is already a relatively easy operation, since a fertile ovary produces tens of thousands of eggs, some of which could be extracted during an abdominal operation. The eggs would be fertilised in laboratory conditions by the husband's spermatozoa, and inserted into the body of the infertile wife, to grow to term in her womb. However, who is to decide on the identity of the woman whose eggs are to be harvested for use? What happens if, as is possible, the eggs are damaged in transit? How can abnormal eggs be detected before implantation? What legal rights would a woman whose eggs had been harvested have over the child which in every biological sense was half hers. Despite these difficulties there is a growing argument in favour of this development. At the moment, with the perfection of birth control methods causing a drop in the number of unwanted children, couples who are infertile have less chance of adopting a baby, and therefore need rely on this form of technology if they want a family. Most scientists would agree that this egg-transfer technique presents technical problems that can be overcome in the next few years. If they are, no one could blame people who desperately wanted children from availing themselves of the opportunity. What happens when having a family involves three people instead of two?

Already there is talk of a product that may have even deeper social significance than that – the production of an artificial womb. No research has yet shown exactly how it could be done, especially since so far no one really understands

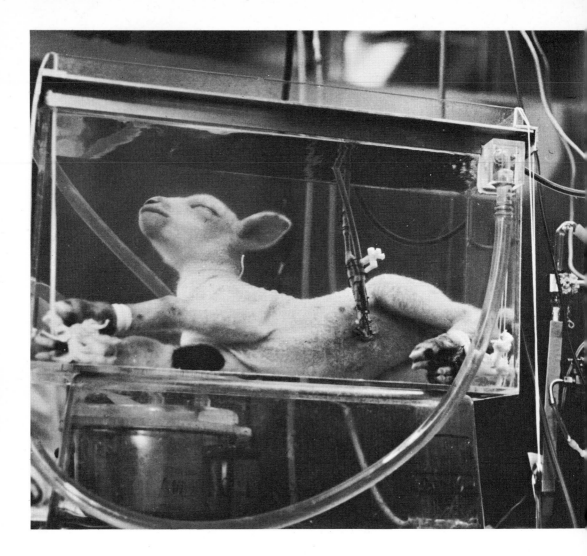

Above The latest experimental artificial womb. An isolated, non-breathing foetal lamb rests comfortably submerged in a tank of synthetic amniotic fluid at the National Heart and Lung Institute, Bethesda, Maryland. The lamb is being kept alive by an 'artificial placenta' system which features a new type of artificial mother's lung – the pint-sized cylindrical objects on the right. So far foetal lambs have lived up to 55 hours in the artificial womb before being 'born'.

Top right While still in the artificial womb the unborn lamb sucks greedily when a finger is placed in its mouth. This is a normal reflex. Human babies often suck their thumbs while still in the womb.

Top left The moment of birth for a test-tube lamb. And it's a girl! After being lifted from its tank of amniotic fluid, which still bubbles in the background, the lamb is carefully wrapped in warm towels. It continues to receive oxygenated blood from the artificial placenta but now it is also receiving pure oxygen via its own lungs from a positive pressure ventilator, on the right. Soon the lamb will be able to breathe normally on its own.

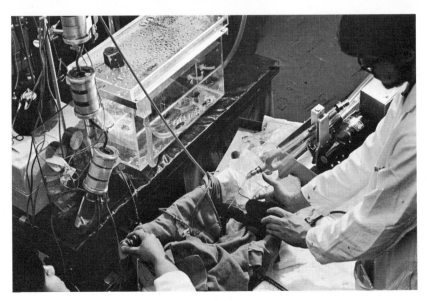

P

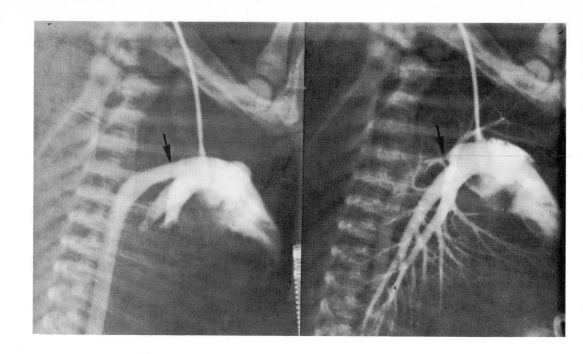

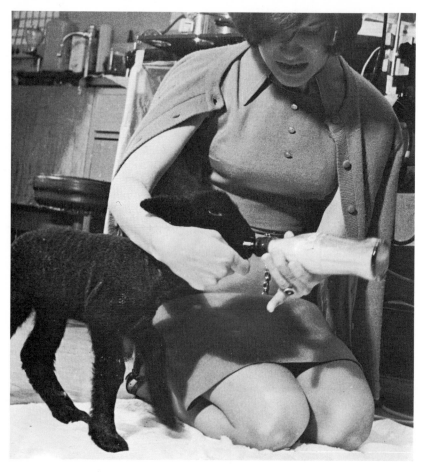

One day old and the lamb – now named Ilya – feeds hungrily. She explored her laboratory home on wobbly legs and a week later was gambolling friskily and nibbling on scientific apparatus.

The first breath of life. These X-rays trace the course of radiopaque dye injected to reveal the dramatic hemodynamic shift — the change from the foetal to the adult form of circulatory pattern — triggered by only slight increases in blood oxygen content. In each photo, the dye is shown injected into a major vein, the superior vena cava, via a plastic tube seen entering from the top in each photo. Travelling with the blood, the dye enters, in turn, the heart's right atrium and right ventricle and is then pumped into the pulmonary artery. The photo on the left, taken during a period of low blood-oxygen tensions that normally prevail during fetal life (40% of normal adult oxygen content), shows that most of the blood is shunted away from the lungs and into the descending aorta through a widely open blood vessel, the *ductus arteriosus* (indicated by arrow), that connects the pulmonary artery to the aorta. The photo on right illustrates another injection of dye in the same submerged lamb foetus only 15 minutes after blood oxygen tension was increased (60% of adult oxygen content). Here, the arrow indicates severe constriction of the ductus arteriosus that reduces bloodflow through it to a mere trickle. Instead, most of the blood courses through the pulmonary artery and its many branches in the lungs, as must occur during the first few gasps of air taken by the newborn at birth.

how the natural womb works. We know that amino-acids, sugars, proteins, vitamins are passed through the placenta membrane to the foetus. We know from experimentation that if those amounts differ in the smallest degree from the norm, the result can be monstrous. The work involved in tracing out in detail every minute of activity in the womb right up to the moment of birth appears almost insurmountable. But as American scientists point out, so did the problems of putting a man on the Moon. Meanwhile work is going on in all the relevant areas, and should it be successful and the artificial womb be produced, the benefits would seem to be considerable. The factory-made womb could be used to further understanding of how to correct faults that can occur during the growth of a foetus in the human womb. From that, doctors might be able to diagnose and treat foetal disease, making sure that every child was born healthy and, even, already immunised. Drugs could be tested by putting them through the artificial womb system to check the effects on a foetus. But all the scientific progress the artificial womb would achieve pales in comparison to the effect it would have on society. For the first time women could avoid the centuries-old business of a 9-month pregnancy. Using the egg-transfer technique, all a woman would have to show for being pregnant would be minor scars where a surgeon made the incision during the extraction of her eggs. And once the age-old confinement and the personal association with the production of a baby was gone, the attitude of a mother and her special relationship with the baby would undoubtedly also change. What then, to echo a previous question, would happen when the birth of a family involved not three, or two, people, but a machine?

Already, one aspect of work on the foetal stage of a human being is causing great excitement. It deals with the fluid that bathes and surrounds us before we are born. The amniotic fluid has been revealed as perhaps the greatest source of information yet discovered about a foetus. The fluid is contained in the uterus of the pregnant mother, and is manufactured primarily from the genito-urinary tract of the foetus and therefore contains foetus cells. If some of the fluid is drawn off through the abdominal wall of the mother and centrifuged, the cells can be separated out and grown in a culture. Examination of the cells has shown so far that no less than 34 different diseases can be identified in the foetus, the most important of which are 24 separate abnormalities which can be detected. Virtually all that can be done at the moment with this knowledge is to advise a mother that her child will be born deformed – and then carry out a therapeutic abortion, provided the mother and the law permit. The very latest work also indicates that the sex of the foetus can be determined. This too, one day, could lead to abortion on the grounds, for instance, that the couple wanted a girl, rather than the boy being carried. Such a thought is horrifying to us today, but may be less so to the next generation. For the climate of opinion is already changing, for example, in regard to the abortion of foetuses carrying deformities like mongolism.

While the search for the fault-free child goes on (leaving aside for the moment the implications of a later search to find ways of making sure that the perfect *number* of fault-free children will be allowed to emerge from whatever womb), technology goes on producing gadgets. The latest developments range right across the spectrum, from care of the newborn, to involvement of the blind in the network of computer terminals that will soon stretch across the country, bringing with them their own special problems for society to adjust to. The inventions are as always haphazard. No one sector of medicine is the subject of a given number of developments in any one year. But all have one thing in common – to aid the doctor in his treatment of the individual.

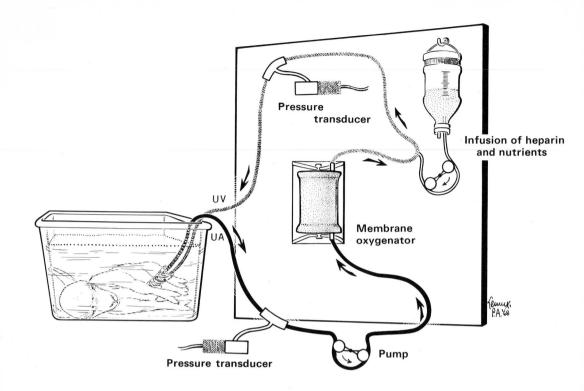

The most vulnerable in need of care and attention is perhaps a newborn baby. Even today in America the mortality rate is 25 in every 1000, much higher than Holland or New Zealand. One of the reasons for this large figure is that there are more high-risk premature births in America than elsewhere. More important, there has been no substantial reduction in the number of newborn deaths there for the last 10 years. This has led to the development of yet another intensive baby-care unit, but one with a difference. The old idea of complex expensive equipment at every baby cubicle has gone. What the researchers at the University of California at San Diego have produced is a facility that is instantly available to every baby in a ward. Gone are the 'in wall' sources of oxygen, electricity, suction and so on that limited the accessibility of the crib. Now everything comes from the roof, on pull-down hoses. This permits floor space to be used more efficiently, since cribs can be positioned so as to receive part of the facility being provided for another infant. Even the lighting in this new unit has been designed to permit the viewing of the truest skin tones, often of vital interest to an observant nurse. Above all, the space-science 'clean room' system of airflow has been adopted. Instead of normal air conditioning, the air flows downwards from the ceiling and out through the floor; this substantially reduces the possibility of cross-infection. And since the baby is no longer at risk from airborne bacteria, another revolutionary idea has been adopted: open incubators. If the entire environment is warm and sterile, and in most cases this is all the baby needs, so the closed plastic box with your tiny baby locked in becomes a thing of the past.

One of the great dangers with premature babies is their tendency to stop breathing without warning. A new device from the British Medical Research Council may solve that problem. It consists of an air-filled mattress divided into ten segments. Each segment is connected to a common manifold and a heated thermistor which is cooled by any flow of air between the mattress segments. A

Plan of artificial womb and placenta. The foetal lamb lies submerged in warm synthetic amniotic fluid *(left)*. It is connected to components of the artificial placenta via specially designed tubes inserted into umbilical arteries and a vein. Blood circuit includes provisions for measuring pressure and flow, for pumping and oxygenating the blood, and for adding nutrients (glucose, amino acids, and vitamins), antibiotics, and an anticoagulant drug (heparin) to oxygenated blood before it re-enters the foetal circulation.

tiny movement of the baby – even its breathing – causes very slight changes in the pressure on various segments of the mattress. Air is forced out to the thermistor, cooling it. This irregular cooling and reheating activates an irregular signal. But if the signal ceases for more than 10 seconds it triggers an alarm signal. A system like this allows one nurse to look after more babies, and frees some of the nursing staff to concentrate on the more serious cases in the ward.

One of the commonest diseases in a child's early years is rubella, or German measles. In most cases it causes only mild distress, and many children often have it without even knowing. The one really dangerous aspect of rubella is having it while you're pregnant. The result can, though not always, be a deformed baby. For this reason it would be desirable for girls under 14 to be vaccinated against the disease. Up to now the vaccine has been grown on animal tissue, usually on cells from monkeys. Animal tissues were chosen for two reasons: the type of monkey used was regarded as a pest; and the cells were difficult and slow to propagate. But now researchers have discovered that there are dangers in using animal cells. Over the last 10 years, more and more viruses have been identified in them, viruses never suspected before. This has led a British Company, Burroughs Wellcome, to develop a vaccine grown entirely on the tissue from human foetus which so far appear to be free from contaminants. A further advantage is that human tissue can be deep-cooled and kept for years. There is no need, as previously, to prepare new animal tissue for every batch of vaccine. It may be that one day in the future you may have a healthy child because you didn't catch German measles during pregnancy – thanks to tissues from a foetus that did not live to be born.

Life-saving box. The baby lies on a unique mattress that will trigger an alarm if it stops breathing. Top right is the respirator that takes over if the child is too weak to breathe on its own.

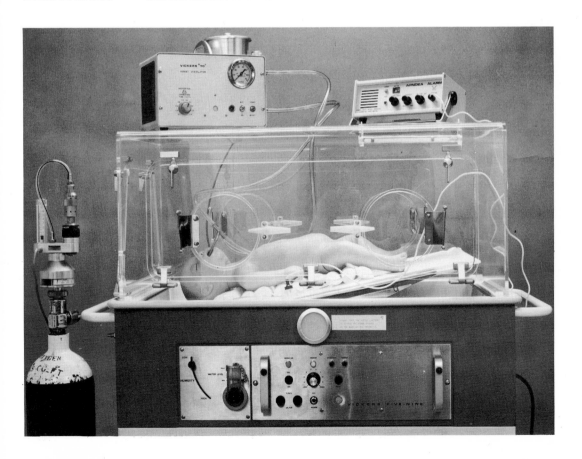

Another common disease throughout the world, affecting some 15 million people, is diabetes. Diabetes results in a build-up of sugar in the urine which occurs when the pancreas, located in the digestive system, fails to produce insulin. Normally most of the body's sugar is produced by the liver, acting on orders from the brain. Insulin aids the passage of sugar from the liver into the bloodstream and, when there is insufficient insulin, blood sugar level falls. The brain senses this and orders the liver to make and excrete more, and after a time 'overflow' occurs. The only place the sugar can go is into the urine, which is why a diabetic's urine is heavily charged with sugar. But the sugar takes with it large quantities of water. In addition, the body turns for the sugar it isn't getting to the body fats, breaking them down. This leads to a build-up of acid by-products in the blood. Extreme symptoms are loss of consciousness, loss of fluid and, from that, lowered blood volume. Shock and death can follow. Up till now the only remedy was to take insulin injections, sometimes as often as every 2 hours. The difficulty with these injections is that individual patients react differently, and absorb the insulin in different degrees. Anyway it is often socially disruptive to have to give yourself a shot every few hours. Now a team at the British Medical Research Council have found a way to increase the effect of one dose from 2 to 24 hours, and may be on the trail to a one-shot-for-life remedy. What they have done is to take the active insulin molecules and attach to them, rather weakly, a group of molecules which transform the insulin from an active to an inactive substance. After injection, the body fluids 'chip away' at the new substance. The weak linkages fall apart and, as they do, release the insulin molecules a little at a time. Tests have been carried out with animals and a few human guinea-pigs, mainly the researchers themselves. So far the results look promising.

Apart from relatively few developments such as these, most of technology's effort continues to provide tools for use in hospitals. This follows a general trend

Glass cylinders contain growing rubella virus for vaccines. The virus lives on cultivated living cells from human foetuses (see page 241).

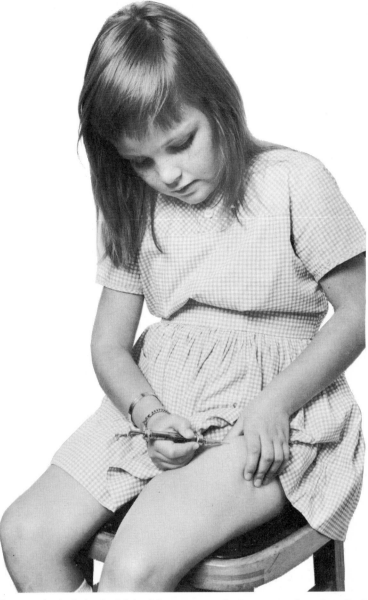

Sally Dutton injects herself every few hours with insulin because she is a diabetic. Now doctors are working to perfect a single insulin injection that will last for life.

in medicine, particularly in the United States, towards the automation of medical treatment. Large, efficient centres like that of the Kaiser Foundation have shown that in many instances the patient doesn't really need to see a qualified doctor. He is tested and diagnosed by medical technicians, who need know only how to operate the machine which plugs into the patient. It is only at the end of the automatic diagnosis sequence that the patient goes, if necessary, to see a properly qualified physician. The advantage is that the doctor has more time to attend to the serious cases that merit his personal attention. Industrial organisations have been quick to see potential profit, and today it is difficult to determine whether, as some doctors think, the gleaming technological marvel that is a modern hospital is there because healing really needs all that equipment, or because the doctors themselves have been taken in by the glitter.

One of the machines that excite some doctors uses the still-mysterious laser, the beam of light that had everybody talking of death rays 10 years ago. This

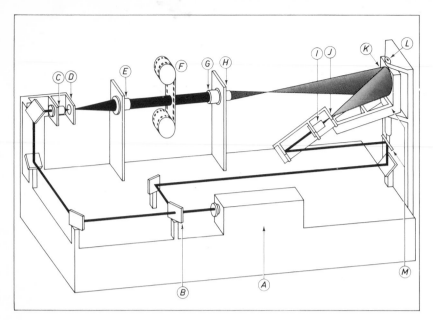

Radiograph Hologram Multiplexer. (A) Laser. (B) Beam split. (C & D) Beam broadened. (E) Beam divergence controlled. (F) Beam passes through X-ray film, picking up image. (G) Beam widened and (H) projected on to (K) hologram plate where it combines with reference beam coming through (I & J) projector. The plate (K) is rotated slightly for every new picture, by mechanical means (M).

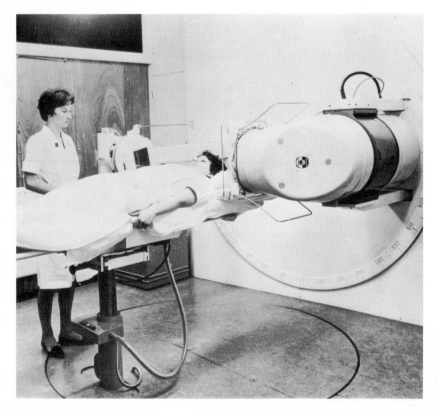

Left A modern X-ray machine of the type used to supply multiple pictures for the holographic multiplexer.

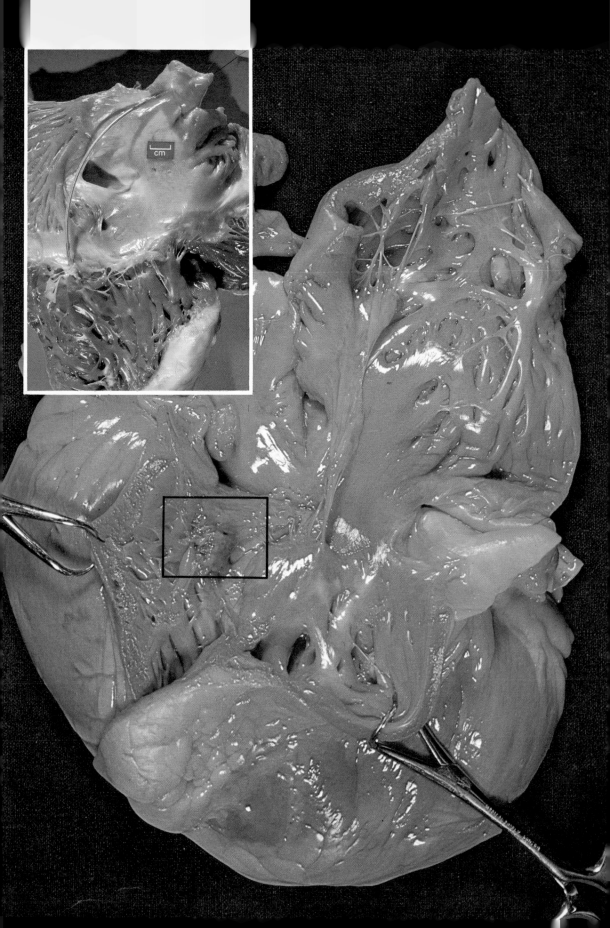

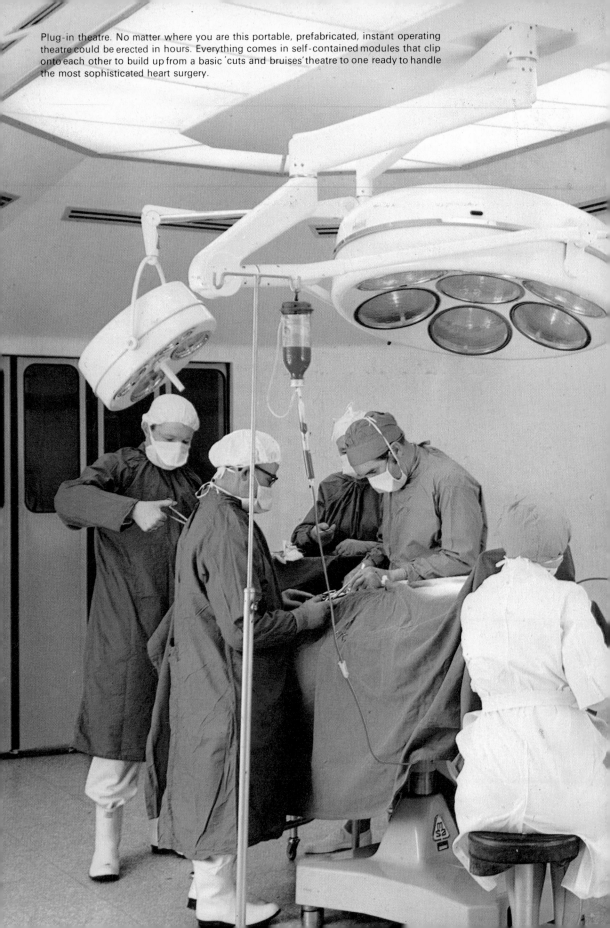

Plug-in theatre. No matter where you are this portable, prefabricated, instant operating theatre could be erected in hours. Everything comes in self-contained modules that clip onto each other to build up from a basic 'cuts and bruises' theatre to one ready to handle the most sophisticated heart surgery.

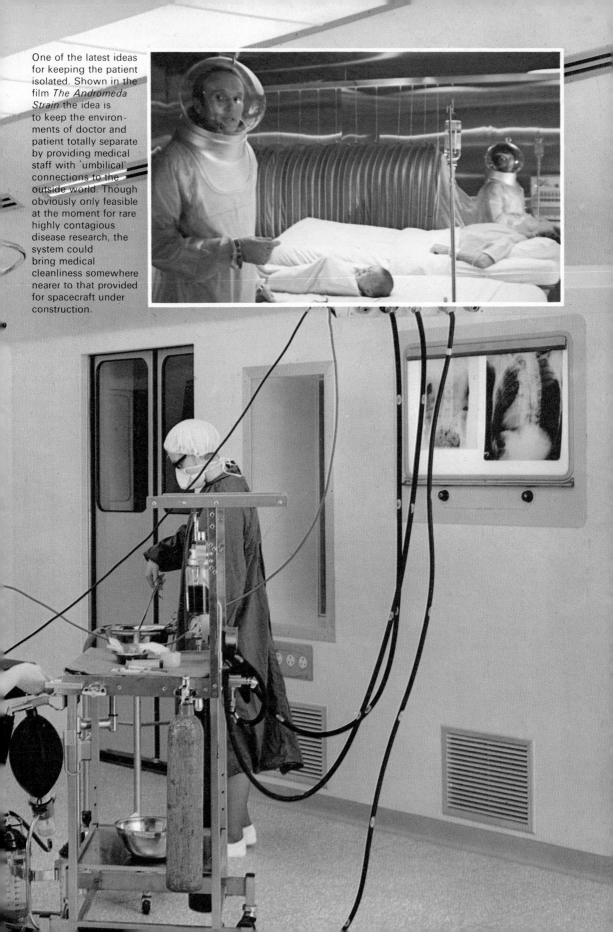

One of the latest ideas for keeping the patient isolated. Shown in the film *The Andromeda Strain* the idea is to keep the environments of doctor and patient totally separate by providing medical staff with 'umbilical' connections to the outside world. Though obviously only feasible at the moment for rare highly contagious disease research, the system could bring medical cleanliness somewhere nearer to that provided for spacecraft under construction.

Left. Colour TV that traces disease. Each smear of colour on the screen indicates varying density of body tissue. A beautiful pattern could mean tragedy for someone. See page 252.

adaptation provides surgeons with a 3-dimensional X-ray picture. Such a view of our interior is of special value to brain surgeons who have to calculate with great precision the site of a potential operation. So far they have had to look at three or more separate pictures, taken from different angles, and work out for themselves where everything is in 3-dimensional terms. The new technique to save them the bother is called 'holographic multiplexing'. It involves splitting a laser beam in half. One part is sent through a radiograph, then matched up, after some complex optics, with the other half. As it passes through the radiograph, it picks up an interference pattern from the X-ray picture and when mated with its other half again, it produces a shadowy 3-dimensional image. So far this is standard holography. The 'multiplex' part consists of taking a series of radio-graphs through an arc of 90° around a patient. These are placed one by one in the split laser beam, and their combined images recorded on a single hologram plate. At each superimposition the image-carrying half beam is put on the plate from the same angle as the original radiograph was taken. The other half of the beam is kept at the same angle to the holograph plate throughout. The result is a series of 3-dimensional pictures presented on the same plate. By moving from one end of the plate to the other the surgeon sees X-ray pictures of the brain 'illuminated' as it were from different angles. He can in effect see the front and back views simultaneously, thus obtaining a remarkable impression of depth and therefore a more exact idea of the spot he is aiming for during the operation.

A considerable amount of advanced technology has concerned itself with the heart, ever since the first heart transplant made headlines some years ago. Recently the US Department of Health Education and Welfare awarded a grant to the University of California at San Diego School of Medicine to take what many regard as the ultimate step in removing from the surgeon's hands the decision to transplant. The plan is to develop a bedside computer to tell a doctor when it is medically *and* ethically feasible to remove an organ for transplant. The system rests on the computer's ability to analyse extensively the donor's brain waves, so as to indicate to the surgeon when the moment of cerebral death is approaching and if necessary to signal him remotely when it has occurred. There would of course be no need for the donor to be in hospital. In a monitored situation at home, the instrumentation could be sent out on a localised radio transmission or by telephone. The announcement of this research project brings the day nearer when advanced medical monitoring will be fed to satellites and thence to transplant centres all over the world, with high-speed transport enabling doctors to reach a donor in time to effect the removal operation.

Meanwhile another computer, designed by the National Aeronautics and Space Administration, will allow doctors to 'walk round' a beating heart, view it from all angles, and decide on whether the heart would make a good transplant or not. The system projects on a computer display screen a 3-dimensional animated cartoon image of any chamber of the heart. Naturally this display can be transmitted to any desired location. First doctors inject into the heart a dye opaque to X-rays. They then take two simultaneous X-ray movie films at 90° to each other, running the film at high speed, because the opaque dye changes the function of the heart, and makes data acceptable for only the first two beats after injection. The entire sequence of the film, frame by frame, is then traced onto a computer's input screen. When the tracing is finished the computer mathematically puts the two right-angle views together, frame by frame, constructing the moving area in a 3-dimensional image formed by crisscross lines of light. As the computer plays the two heart beats back over and over again, doctors can study the movement and identify damaged areas of the heart wall, any gross malfunction, or holes between heart chambers. Combined with blood-flow measurements this technique can also help identify inefficient pumping activity.

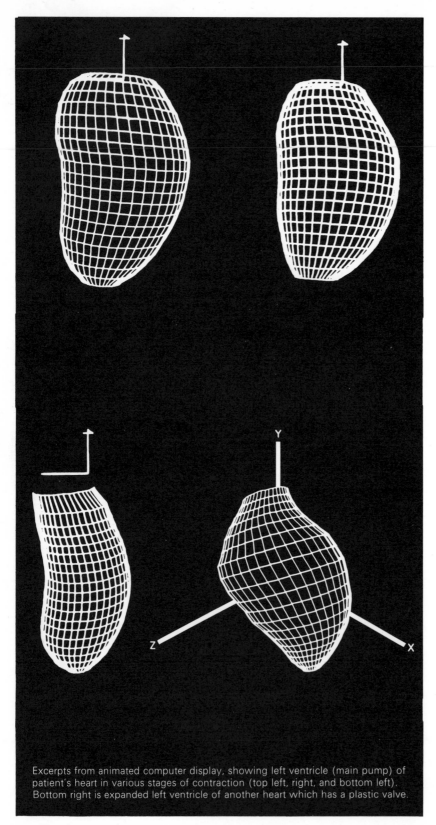

Cardiologists perform a heart X-ray, the first stage in diagnosis of heart disease using a computer. A catheter is inserted into the patient's arteries to gather information for the computer to create a 2-dimensional animated cartoon of her beating heart.

Doctors confer. The display shows the left ventricle (main pump) of a patient's diseased heart as it beats. With the system, doctors will be able to stop and start the heart, look at it from various angles, run it in slow motion and make other studies.

Excerpts from animated computer display, showing left ventricle (main pump) of patient's heart in various stages of contraction (top left, right, and bottom left). Bottom right is expanded left ventricle of another heart which has a plastic valve.

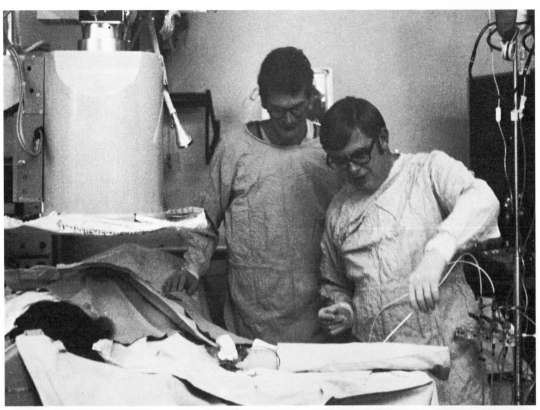

One of the systems which might be coupled with computer animation is the result of a project going on at the Massachusetts Institute of Technology into the possibilities of *hearing* anomalies in blood-flow. This is of particular importance in studying cases of arteriosclerosis, or hardening of the arteries. The instrument being developed is called a phono-angiograph. It is, in essence, a tiny pressure-sensitive microphone. Placed directly over the suspect artery, it picks up the rushing sound of the blood as it passes along the blood-vessel. At a point where fatty substances have caused a thickening of the artery wall, often a precursor to heart attacks if the artery is feeding the heart, the blood has to pass through a constricted space. Immediately afterwards turbulence develops, as the blood rushes into the wider after-section of the artery. It is this turbulence that causes the particular noise the microphone is listening for. The signal from the microphone is picked up, amplified, and played into a sound analyser of the type used to make voice prints of human speech. The end-product is a strip chart with tracings showing the pattern or spectrum of the sound in terms of intensities and frequencies. With this chart the engineers and doctors can relate the sounds the blood is making with predicted patterns for different types and forms of arterial constriction, and thus arrive at a diagnosis without puncturing the skin of the patient.

With all this welter of super-sophisticated hardware going into hospitals, the technologists have sensed a good market for more products. One British company has produced an instant, modularised, plug-in operating theatre. It doesn't matter where you are, as long as you have water, steam and electricity supplies on tap. The obvious aim is to appeal to the poorer hospital, or the well-heeled group of doctors who are too far from one of the big hospitals to make efficient use of it. The walls and ceilings are made of galvanised steel frames with plastic panelling. The design is modular, so that any one of six sizes of theatre can be built, according to the kind of surgery needed. Built-in air-conditioning changes the air twenty times an hour, and all essential supplies come in retractable tubing swinging down from a central facilities-beam. Everything comes in standard sizes for ease of replacement or addition.

Whether you have your operation in a steel theatre in the backwoods or a giant medical centre in the city, the chances are that early analysis of your tissue, your bones, or indeed virtually anything that needs microscopic examination, will be carried out in the near future on a development of the television tube. This is a new colour television analyser, at present undergoing trials at one of Britain's major hospitals. The device scans black and white photographs by passing a beam of light through a transparency. As the beam comes out the other side, degraded by the level of density it has passed through, it registers on a light-sensitive cell. This gives off an electrical signal relative in strength to the amount of light, and therefore to the amount of density in the transparency. That signal is passed to magnetic tape for storage. When the tape is played into a modified colour TV set the signals are fed to one of seven colour-producing sources, relative to the original density on the transparency. So for instance black to dark grey could be presented as blue, dark grey to medium grey as red, and so on through the seven-tone colour scale. Besides giving beautiful pictures, clearly showing different gradations of substance density, the information can also be passed to a computer for analysis. That in turn can be compared with a pre-programmed memory store, and the comparisons will show any differences from a previous state, or any deviation from normal densities in a healthy body of the substance under study.

Once the surgeon has analysed what's wrong and starts work, show no surprise if he appears to be talking to himself. He may simply be lecturing to a group of students watching the operation on TV in a distant room – or dictating notes for

At the touch of a button all the nurse needs to know about a patient's condition and treatment is flashed onto the TV screen. The information comes from a central computer. In some hospitals this system has already replaced hand-written notes and charts.

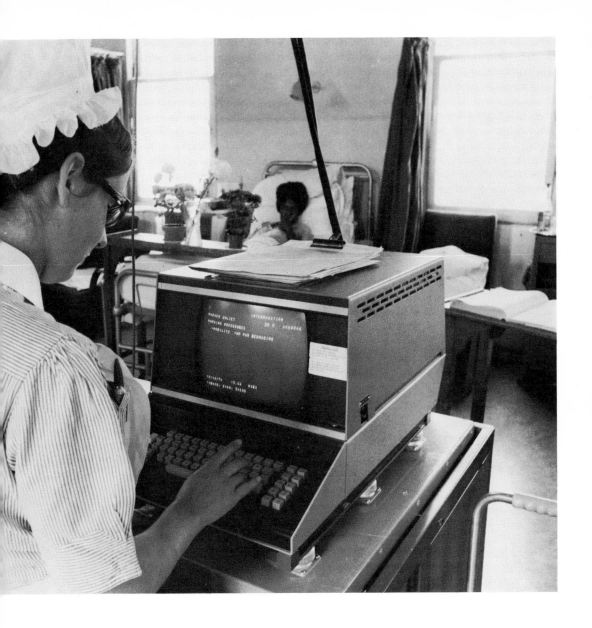

consideration afterwards. The microphone will be under the surgical mask covering his nose and mouth. A tiny transistorised affair, it is a development of the British Ministry of Aviation Supply. Or it might be that the surgeon is only ordering the medical music to be turned on.

This 'musak' is no different from the stuff you hear every day in shops, hotel lobbies, restaurants. Its main purpose is to soothe you. In the case of the experiments at Stockholm's Maternity Hospital, it could soothe you into a state of anaesthesia. Dr Arne Mellgren, the head of the Psychomatic Medicine Department at the hospital, has discovered that sound can be an effective pain-killer for anything a local practitioner might do in his surgery. The only thing he dare not work on while it is being used is your ears. Dr Mellgren has found it particularly effective during minor gynaecological operations, especially in the case of patients who, because of cardiac or other conditions, are limited in the amount of conventional anaesthesia they can safely accept. Music is mainly used for lower pain-threshold operations. The patient wears headphones which give the same level of volume that you would hear in a very small room with a very big hi-fi set. As the operation proceeds the patient has control of the volume level, 'to give her something to manipulate and occupy her mind', says Dr Mellgren. The maximum noise level is preset to 3 decibels below the point where the noise itself might cause pain. Most people can stand maximum volume for 10 minutes, the time an average minor operation takes. Analysis has revealed that while Mozart and Vivaldi are firm favourites, 'most mothers prefer light music'. But whatever they prefer, there seems to be little doubt that the system works. The anaesthesia bills in Mellgren's clinic are down by 25%. Ironically, by the time this system is in use to any extent our pain threshold for noise may have risen drastically. Over the last 50 years the general noise level in industrialised societies has increased to the point where the ear itself may be losing its sensitivity. Today we may, as one scientist pointed out recently, be as much as 10% less aware of sounds than people were at the end of the last century. And by the year 2000, unless stringent noise-pollution laws are passed, we may only be able to hear *half* what we can hear now.

At the other end of the cost scale from portable, automatic tape-recorders playing anaesthetic music is the new cobalt cathetron. Developed by Atomic Energy of Canada, and installed at the University Hospital of San Diego, California, this device removes hospital staff from the danger of contracting one of the latest and most awesome man-made diseases: radiation poisoning. In the past, radium treatment of tumours involved the insertion of the radioactive source, and therefore a prolonged and potentially dangerous exposure for the medical staff. Attempts were made to minimise the danger by offloading the radium into protective sheaths as it was extracted from the patient; but it was a lengthy business. Now the whole thing can be done by remote control. The sources are three high-intensity cobalt 60 pellets inside a protective lead barrel. Catheters (tiny plastic tubes) are inserted into the patient through the body's natural openings to the area which is to be treated. These tubes are clamped into position, connected to cables, and the operators leave the room. The pellets are then inserted automatically into the catheters and pushed up inside the tubes. Once in, the doctor at a console in another room can manoeuvre them into position along the tube, or slide them up and down to irradiate a wide area inside the body. The speed with which this operation is performed, since it needs no insertion surgery, means that patients can come for treatment on an out-patient basis and receive as little as 2 minutes' irradiation. Those working on the San Diego trials hope to develop techniques that will enable radiation treatment to be as quick and easy as a visit to have your tonsils looked at.

If an out-patient visit *should* turn into a stay in hospital, tomorrow's patient can be sure that considerable research has gone into the object with which he will be most intimately involved – his bed. Every year, new bed designs are produced. Strangely enough, while they are eager to install some of the more astronomically expensive status symbols of medical treatment, many hospitals show little interest in new beds. The latest one uses techniques borrowed from NASA. During preparations for a space shot, and in particular the building or assembling of a spacecraft or a piece of equipment that will touch the surface of another planet, the greatest care has to be taken to ensure that the object is sterile, that it carries no bacteria. This work goes on in what are modestly called 'clean rooms'. It used to be said that compared with clean rooms, most hospital theatres were garbage heaps. This will soon no longer be true. The newest intensive care unit at the San Diego University Hospital uses the NASA 'laminar flow' principle to surround its dangerously sick or very vulnerable patients with a protective curtain of air. Special isolation canopies over each bed pass over the patient air which has been sterilised with ultra-violet rays of high intensity. On the way through the airstream picks up micro-organisms and bacteria from the patient and carries it through ducts under the beds to be sterilised and then recirculated again. Although there are six highly contagious patients in the same room, the 'air curtains' will protect one from the other, and even allow the nurses to pass from one patient to another if necessary. The placing of the canopies in the air above the bed also means that the floor space remains free of clutter, which would make the simple downflow of air difficult to achieve without some measure of turbulence, and therefore increase the danger of cross-infection.

Long-term patients pose a different type of problem. In their case the normal bed can be the source of intense irritation and even pain, and patients have to be turned frequently if bed sores are not to develop. A new bed from another aerospace institute, the British Royal Aircraft Establishment, provides the answer. The mattress contains a dry incompressible fluid, contained within a woven cover through which water vapour can readily permeate. This 'fluid' consists of millions of very small beads: as you lie or shift on the mattress, the beads pack themselves in round the shape of your body and support it. When you move again the beads redistribute themselves to give the best cushion for your new position. This kind of bed is of considerable use in geriatric cases and in the treatment of the later stages of burns. Because these patients are extremely sensitive to pressure, they benefit from the fact that, since the beads distribute themselves to fill every curve of the body, they provide uniform support and therefore eliminate specific pressure points.

For the patient who has the most agonising time in his bed, the burn victim, another British development should help. In the early stages of the treatment of burns, the skin is raw and 'weeping', as the body gives off protective fluid over the burnt area. The result is that the patient finds it very painful indeed to lie down because of the contact with the bed. In addition the 'weeping' also needs constant drying off. The new 'Hoverbed' deals with both problems. It is formed of a number of pockets, filled with air from a pump underneath the bed. The pockets themselves are flexible and open down the centre of the bed, so that the airflow escapes upwards after filling them. When a body is placed on the bed and the resultant pressure downwards equals the upward pressure of the air, the pockets fall away round the silhouette of the body, leaving it supported on a cushion of air. The temperature and humidity of the air can be controlled according to the needs of the patient. The advantages for severe burns victims are clear, but the story of the development of this bed illustrates the fascination the gleaming machine, rather than the more mundane object, has for hospitals. Suffice it to say that the Hoverbed, and its successor, the low-loss air bed, working on the same principle, have been 'in development' for 10 years. The authors personally examined the bed over 5 years ago, and it was virtually the same as this production model. In all that time it would appear that lack of financial backing prevented the bed from coming on to the market sooner. Burn patients please note.

As technology provides the means for treating more and more patients, and casts an ever wider net in the search for new problems to solve, the one limitation which does remain constant is the ability of nurses to handle the increasing numbers of cases. Already in most parts of the world there is a severe shortage of trained nurses. This may be because in many countries they happen to be paid a salary that forces them to live like paupers, and are often treated with all the courtesies a delinquent army private might expect. Be that as it may, technology has found in their problems another market. The new device is a bedside communications system which enables nurses to talk to patients without actually

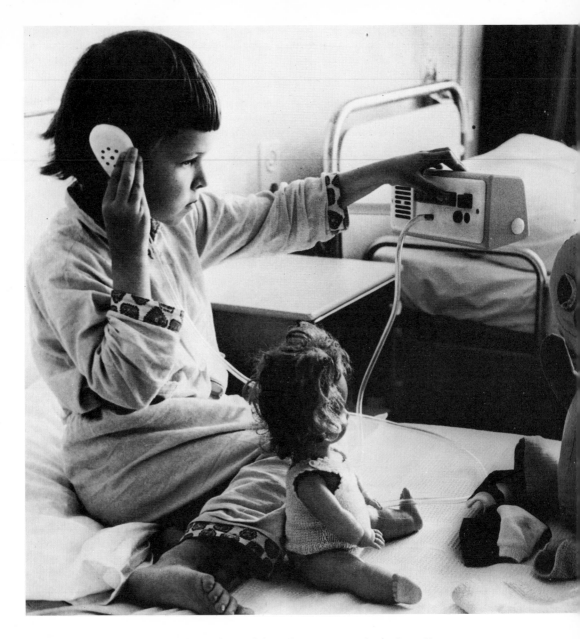

having to be at the bedside. This can be useful on the many occasions when all the patient needs is somebody to talk to. The system consists of a central switchboard, manned by a telephone operator, which visually displays the location of every nurse. As a nurse enters a room, she touches a switch on the wall, activating a light on the console and at the same time opening a speech circuit between her and the operator. A similar switch unit is mounted on a movable arm fixed to the wall above each patient's bed. On the unit are controls for calling the central console, to talk to other patients, to receive radio programmes, or to signal an emergency if need be. In a situation where help is needed the operator can identify and warn the nearest nurse to deal with the problem. The unit on the patient's bed will also pick up breathing noise. In cases where there might be respiratory trouble, the operator can selectively monitor without going near the patient.

While the communications system is an excellent one, its very development points up one of the main indications of what hospitals of the future will be like. The trend is very clearly towards greatly increased automation. 'This,' say the hospitals, 'is what we are all about. We are here to cure and discharge people, the quicker the better.' But research increasingly shows that how people feel and react to being in hospital has a considerable effect on how quickly they recover. One eminent British neurosurgeon has even stated that the patient's emotional and psychological condition before undergoing an operation could affect the outcome to the point of success or failure. There is considerable danger that, in a desire to streamline, to make themselves more efficient, the hospitals of tomorrow may find themselves with a new and incurable disease – the patient, surrounded by racks of equipment, communicating with machines, even consulting a computerised diagnostician, may succumb to loneliness and alienation. Once again, as technology solves one problem, it creates another.

One of the hardest jobs for a nurse is giving a patient a bath. Heaving a full-grown, incapacitated man can make a small nurse's life very tiresome. It is with this in mind that the Swedes have produced a human-washing system to take all the backache out of bath night. It's an integrated transport and bath system which allows the patient to remain on a stretcher from bedside to underwater and back. The stretcher is made of fibreglass. A hydraulic foot control adjusts the height to that of the bed, and the patient is gently slid on. At the bath the trolley supporting the stretcher is manoeuvred so that the stretcher is resting on the built-in hydraulic lift table within the bath. The stretcher trolley is disengaged, and the patient is lowered into the water, stretcher and all. The reverse manoeuvre takes the patient back to bed. The bath itself is of steel and has a control panel including thermostatic mixers, hand shower and controls for drainage and disinfection. The only thing the patient has to do is say if the water is too hot. And this raises another problem.

One of the difficulties of which modern medicine is becoming increasingly aware is the frequency with which the doctor has to rely on you, the patient, for a description of how you feel. This may be of little importance if what you are haltingly describing is a simple headache. At the other end of the scale it may be vital that for correct diagnosis and treatment the doctor receives the kind of subjective information you are incapable of giving. You may not know the right words, your inarticulateness may be a dangerous setback to correct diagnosis. This is, for instance, true of people with artificial limbs or plastic replacements like a thigh joint. In the past the doctor has had to wait until the patient can explain how his new leg isn't working properly, but now a British research organisation has provided doctors with this information without the patient uttering a word. They have developed a new pair of shoes which analyse how you walk and can tell a doctor instantly whether your new hip socket, or leg, or knee cap, is going to give trouble. The basic instrument in the shoe is a load-measuring transducer – flexible metal filigree sheets sandwiched between layers of sponge rubber in the sole of the shoe. As load is applied, the space between the metal sheets changes, and their electrical reaction one to the other alters. These changes are passed to a 'transmitter' in the heel of the shoe, and the signals are fed into an amplifier and converted into displays which show how much weight you are putting on your foot, and in what way. This information can be extrapolated by the doctor to tell him how these movements, or to what extent any limp present, will place further loads on parts of the leg all the way up to the hip. With this information there is no need for you to wait until you feel the pain before coming back to the doctor to complain. He will know before you leave the building!

Whatever the hospital of the future looks like, whatever the developments in

Q

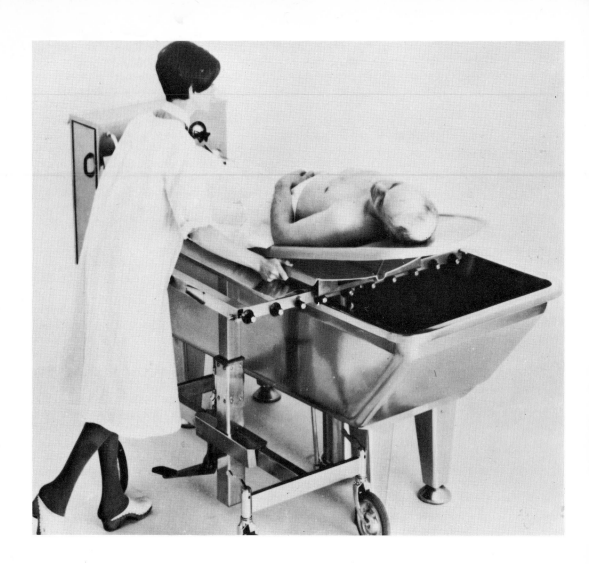

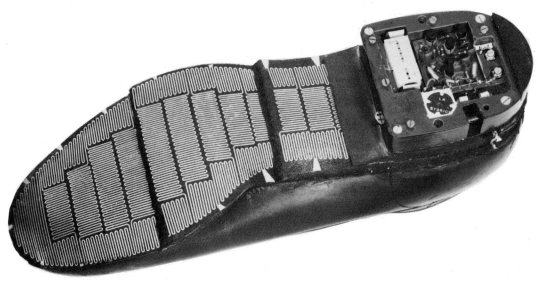

medical knowledge over the next decades, the one certain thing is that we stand a better chance of living longer than ever before. This is, of course, the ultimate aim of medicine. Every doctor must do his best, use whatever instrument or technique available to ensure that his patient lives as long as possible. Medical technology is far too big an industry to slow down, let alone stop the production of new devices to aid the doctor in his work. But it is that very concentration on the individual that has brought us now to the threshold of what may be the greatest social changes in history. The success in fighting what used to be the mass killers, pneumonia, TB, malnutrition, and so on, has produced an accelerative increase in the number of old people alive and well in our society. As the number increases over the coming decades, we will have to be prepared to accept a new set of values in which the minority, those between 20 and 40, will be working to support the majority, children and old people. Whether or not we can manage to do so without rethinking the entire concept of the individual's relative value to society remains to be seen. Opinion in general is that we will be faced with one of two ways out: either to leave things the way they are, and demand considerably more in the way of taxes to support the old and unproductive; or by the end of the century instil in the members of our society the concept that euthanasia is an acceptable alternative.

The situation looks like being further complicated by the advances in the field of birth-control. Most drug researchers are confident that a perfectly safe, chemical birth-control agent will be developed, involving either long-term contraceptive action or virtually instant abortion of any fertilised eggs once a month. The development of the pill and other contraceptive devices over the last few years has already brought about changes in behaviour that give some indication of what the effects of foolproof long-term contraceptive means might be. Whether or not, as some opponents of contraception believe, it will lead to a disastrous drop in the moral standards of the population remains to be seen. What it will undoubtedly do is remove the close association between intercourse and reproduction. That in itself will call into question a moral attitude to sex that our society has lived with for centuries. There are many who believe it will lead to the break-up of the family unit, since it will no longer be considered wrong, or dangerous, to have extra-marital intercourse.

But even if we adapt to these changes in our moral attitudes, and learn to live with an extremely high level of taxation of the relatively young and productive in order to feed, house and clothe the old and unproductive, the wider implications of perfect contraception are staggering. The number of children being born will fall. Perhaps the numbers will reach the goal desired by so many of the world's population experts: a 'zero growth rate' by the end of the century. If that happens, and the number of children being born does not exceed the number of adults dying, then medicine will have contributed to the greatest change in the structure of our society since the beginning of recorded history. With less people coming into the world and less leaving it, the average age of the population will increase drastically. Anyone reading this book who is over 30 may be a member of the last generation to be born into a young world. Throughout the centuries, disease and a rising birth rate kept our society young. What will happen when, because of the efforts of medical science, it ages overnight?

11   What we want is ...

Some while ago a television reporter we know went into the streets to find out what people thought was the most significant and far-reaching scientific and technological advance of the century. Would it be Rutherford's achievement in splitting the atom? Or the ability of Man to travel through deep space? Or the discovery of antibiotics?

The reporter stopped a middle-aged woman and, inching her carefully before the camera, put the question. The lady said she would have to think carefully because there had been so many important advances recently. The reporter prompted her a little. But, no, it was not atom-powered electricity-generating stations. Neither was it supersonic travel. It certainly was not Man in space. And she dismissed the invention of plastics with a wave of the hand.

Then her face lit up. There was no doubt about it. The most important scientific and technological breakthrough this century was the invention of disposable sandpaper sheets for the bottom of bird cages. The reporter was incredulous. Said the lady, reprovingly, 'Have *you* ever tried to clean out dried-up parrot droppings?'

The point of this story is that most people in the advanced Western countries are only too happy to latch on to scientific and technical innovations when they fully understand *how* they will make life easier and more pleasant. In that respect, the simpler the development the better. And the cheaper, too.

It may be puzzling and sad for the technologist when there is hardly a flicker of public interest in the creation of brilliant new fluid logic devices despite their far-reaching implications. It may sadden him even more to see the excitement on the face of his own wife at the prospect of a relatively simple piece of technology – say, a new oven that cleans itself at the touch of a button. Yet, even for a technologist's wife, *understanding* must come before appreciation. After all, do you understand the concepts of fluid logic? Have you even heard of it? (See the previous volume of *Tomorrow's World*, published in 1970, page 180.)

As science and technology become more sophisticated, the spin-off for the consumer is inevitably going to be more complicated too. That is why we are faced today with the promise of computers in the kitchen, portable teletype machines by the bed, and personalised plastic credit cards in our pockets instead of old-fashioned money. But given the choice, is that what people are really looking forward to?

If you could order an inventor to work on an important new product, or persuade a scientist to concentrate on a problem of your own choosing, what would you tell him to do? In the late spring of 1971 we put the question before many millions in the form of a series of newspaper advertisements.

So, as inventors and their commercial backers invest capital to influence our futures, here is a selection of some of the things their potential customers say they *really* want. And since modern women are well aware that science and technology are capable of providing them with the hardware and know-how that lead to better living, let them have the first say.

'Medical scientists should work to find a new drug that allows a mother to give birth to her baby while fully awake – but with absolutely no pain or fear,' says

Mrs A. S. of Winnipeg, Manitoba. 'It should be the kind of preparation that does not harm the baby in any way. I told this to my doctor once who said that it was better for a woman to experience some pain. Needless to say, my doctor is a man.'

'I want an automatic vacuum cleaner which, when placed in a closed room, removes all loose dust without moving; an instant thaw machine for deep-frozen foods; a ground-level smoke-consumer to replace the factory chimney; combination locks on front doors to replace house keys; solid tyres for vehicles – especially childrens' bicycles; a hood containing dry-cleaning powder for use with home hair dryers for between-shampoo grooming.' Mrs J. B., Mitcham, Surrey.

Swear stopper
'Someone should invent a magic eye that fits to the front of cars and automatically dips the main headlights when another car approaches at night. It would stop my husband swearing – and I don't blame him in the least.' Mrs B. McD., Middlesex.

'I would like someone to invent my very own robot, so that I could have a conversation with it when I am lonely. It should be able to help with my chores and to watch the children when they want to play outside. My robot would be a sort of metallic au-pair and it would give me the best of all worlds – a friend, a helper, a nanny and, unlike most dishy au-pairs, it would never turn my husband's head!' Mrs C. A. T., London NW.

'I'm amazed that no one has yet come up with a four-legged chicken. It would save a lot of argument in our house.' Mrs M.M., Nottingham.

'If we can fly to the Moon, why can't scientists create a harmless bacteria that added to packaged foods puts back the old-fashioned taste we used to have?' Mrs R. F. W., Vancouver.

Pimple power
'Doctors should wait no longer to produce a sure-fire one-dose pill that removes adolescent spots for ever.' Miss S. U., White Plains, NY.

'I wish someone would invent an "umbrella" attachment for outdoor rotating clothes dryers. At the first sign of rain it would automatically erect itself to keep the clothes dry.' Mrs K. V., Oxfordshire.

'This may sound trivial but I bet a lot of people would want it. What about creating a preparation that when spread on a lawn stops the grass growing beyond a certain height?' Mrs M. W., Yorkshire.

'What I would like is a draught tracer!' Mrs D. K., Kent.

Auto-translator
'I would immediately order an inventor to concentrate on building a computer that can translate the spoken word into foreign languages. I am convinced the main obstacle to peaceful co-existence is language, so this seems to me to be an important invention.' C. H., Netherlands.

'What I need is a means of deterring flies, wasps and mosquitoes at open doors and windows. Is there no scope here for some kind of death-ray?' Mrs D. B., Hertfordshire.

'Since energy is indestructible it should be possible in some way to hear the actual voices of the great people of the past. I would like to hear what King Henry V *actually* said before Agincourt!' Miss V. A. L. (SRN), Sussex.

'Vacuum cleaners are out of date. There has got to be a scientist somewhere working on a machine that will attract dirt, dust, fluff, and turn it into a disposable pill.' Mrs R. P., St Boniface, Manitoba.

Miracle metal
'I am waiting for an inventor to make a metal grill for a gas cooker that does not discolour as soon as it gets hot.' Mrs A. McD., Dundee.

'I would like a gadget that lifts off the lids of cream jars and jam pots so that they are in a fit state to be replaced afterwards.' Mrs J. D. A., Norwich.

'When is somebody going to market an aerosol can that puts back the healthy, sweet smell of natural under-arm odour?' Mrs I. C., Denver.

'Why can't engineers work out a way to fix parachutes to the cabin sections of large passenger aeroplanes so that a crashing aircraft need not fall uncontrolled out of the sky? If they can do it with an Apollo space capsule, why not, eventually, a jumbo jet?' Miss E. R., Edinburgh.

Sticky millionaire
'When scientists have the time they should find the formula for an all-purpose pliable material to fix anything from a cracked furnace to a broken coffee cup. It must be invisible when dry. The clever man who does this will make a million.' Mrs D. P., Vancouver.

'I believe I have discovered why most great chefs are men. It's because their taste buds are not confused by the flavour from lipstick. Can't somebody make a flavourless lipstick for women? I hate the taste of them all!' Mrs F. C., Munich.

'It should be possible to design and build a personal hover-shelter for use in the event of earthquakes. At the touch of an emergency button it would rise above the heaving earth and hover until things had settled down. It would have a reinforced roof to protect the occupants from falling debris and it would be water-tight so that it could float in the event of tidal waves. It should be made available at once to all hospitals and schools in the San Francisco area.' Mrs L. de K., San Diego, California.

'Without doubt the invention that would bring the greatest benefit to the world would be a domestic fuel cell,' says Mr R. W. of Staffordshire. Perhaps it is because the head of the household is usually faced with the bills that many of the men who came up with ideas for this chapter thought that cheaper and more efficient ways of producing power should be the first concern of inventors and scientists. Mr A. W. W. of Essex had a particularly European view: 'I suggest that inventors should seriously apply their minds to inventing and making public a method of transmuting water into engine fuel by the addition of a simple chemical. It would utilise the basic oxygen and hydrogen as a propulsive power, thus replacing petrol. Cheapness would be no small boon, and reliance on the whims of oil-producing countries would be unnecessary and existing political difficulties would end.' He suspects there would be considerable opposition to such an invention, though. 'But economic considerations must surely give way in the light of the growing level of world pollution.'

'Scientists should concentrate on improving the quality of the air we breathe and the water we drink. They could do this by perfecting an electric car engine and by finding a foolproof way of removing dangerous levels of lead from our drinking water supplies.' Mr J. P. T., Cardiff.

'It is high time technologists produced a small, compact, silent electricity generator capable of supplying the power needs of the average house. Then the power companies would have to reduce their charges or lose business.' Mr R. T. B., Manchester.

Powder plates
'The time has come for a total rethink of the existing domestic washing-up system. It should be possible to create a machine which would manufacture food utensils from a stock of powder. At the end of the meal the same machine would evaporate these containers and break down the raw material into an acceptable fertiliser for absorption by the soil.' Mr L. H., London W5.

'Scientists should be working on a "love drug". When sprayed from an aerosol it would calm rioting mobs and political demonstrators without doing them any physical damage.' Mr A. E. R., New York.

Hot pants

'If they put their minds to it, inventors and scientists could produce a clothing material (for pyjamas especially) that is always warm to the touch no matter how cold the surrounding air.' Mr A. P. W., Caernarvonshire.

'I would like an atomic-powered car. Well, why not?' Mr J. L., London SE.

'I suggest that the scientist-inventor who discovers a nuclear car battery to drive vehicles without consuming oil, petrol *or* electricity and avoids all air pollution, should be rewarded with fabulous wealth – and a Peerage!' Mrs R. S., Cheshire.

'Who is going to invent a reliable, miniaturised car telephone that is cheap?' Mr P. A. L., Hamburg.

Wasted water

'One of these days someone must invent a domestic filter and a pump so that used water can be returned to the hot water tank. It would save billions of gallons a year.' Mr B. R., Paris.

'How about a 5-minute washing centre for people. You would go in, be stripped, soaped, washed, dried, shampooed and blow-waved, manicured, pedicured and deodorised. At the same time your clothes would be dry cleaned, shoes polished, socks mended. At the end of the line you would emerge like a new person.' Mr E. J., Yorkshire.

'I suggest that scientists should work on a device that is worn in the mouth to allow a person to breathe oxygen diffused from water. Much better than aqua-lungs. Much safer than a lifebelt.' Mr M. M., Lancashire.

'Please instruct an inventor to produce a printing ink that actually stays *on* a newspaper. The ink in use today blackens hands, clothes, bedsheets – and anything else it comes in contact with.' Mr C. C. S., Surrey.

'I would order an inventor to make equipment for cars designed to throw out a small force-field while the car is in motion. Such an invention would eliminate traffic accidents as vehicles would never be able to make contact with one another in potential accident situations. When he had done that I would get the inventor to perfect a "Dicta-Riter" – a machine that automatically types out the spoken word without spelling mistakes. Then he could turn his attention to an anti-gravity device that would be fitted to your shoes so that police and firemen could climb the sides of buildings in an emergency. After all that, something simple. The inventor should be able to design a ball-point pen that writes on greasy paper.' Mr B. C. G., Kent.

'I suggest that scientists make home cine films that develop themselves as soon as they are taken. Then when we get fed up with that, they can make a cine film you can wipe clean (like magnetic tape) and use again.' Mr O. V., San Antonio, Texas.

Misty eyes

'It seems ridiculous that glass is so vulnerable to atmospheric conditions. Why can't someone produce a glass which is not subject to vaporisation, conden-sation, and moisture in general? Then there would be an end to steamed-up windows.' Mr H. E. H., Surrey.

'Scientists should be working on a generator that emits silent and harmless energy waves that destroy fog and smog particles.' Mr G. T., San Francisco.

'I would order an inventor to make a song-making machine. The trouble is these days the words don't make sense. Perhaps a machine would be more successful.' Mr T. K., London E3.

Perhaps it is not surprising that more than half the hundreds of suggestions put forward by people from twelve countries[1] were directed at medical scientists. The urgent need for better cancer and heart disease treatments headed the list. The lady from Washington D.C. who thought it should be possible to make a liquid solution that would flush out blood vessels 'like flushing out a sink' echoed the man from Newcastle-upon-Tyne who wanted doctors to invent a 'pull-thru' for unblocking coronary arteries.

An Irishman from Belfast felt that transplant surgeons should be improving their techniques so that larynxes of famous opera singers could be preserved to make music long after the rest of them had perished. Pretty far-fetched, perhaps, but not so unlikely as the demand put forward by a market gardener in Devon. He felt it should be possible one day for doctors to find a way to 'fertilise human brain cells so that they can compete with computers before they take over the world!'

Yet one day some of these lay ideas directed at inventors and scientists alike may be feasible. A senior citizen in Aberdeen wants doctors to develop a biological glue that will repair broken bones 'especially in old people'. Another old person from London wants a 'simple' drug to improve the fading memories of old folk. A schoolboy in Wales wants a pill that will make it easier to 'learn anything, especially arithmetic'. And there were dozens of requests for scientists to discover a safe substitute for tobacco.

One is tempted to believe that many people are convinced that, provided the money for research is made available, there is practically no biological problem that cannot be solved by medical scientists. The doctors themselves might not take such an optimistic view, but it must hearten them to know how much faith the layman has in their talents.

So here is a selection of ideas to influence future lines of medical research. A single success story among them will confirm the widely held view that medical science can achieve almost any breakthrough.

'Doctors must concentrate on a cure for acute depression. In the meantime why doesn't a psychologist create a machine on which very complex games can be played? It would keep depressed patients fully occupied and absorbed – something that wool rugs and basket making can never do.' Miss P. F., Birmingham.

'Scientists should find a cure for Huntington's disease, which is a condition passed on from generation to generation. Hardly any work is being done at the moment.' Mr B. S., Yorkshire.

'I would like an inventor to devise callipers which can be adjusted by a child. At the moment all the available designs need adult attention before they can be adjusted from standing to sitting positions.' Mrs J. S., Warwickshire.

[1]Belgium, Canada, England, France, Germany, Holland, Ireland, Italy, Scotland, Switzerland, United States, Wales.

'I want doctors to find a way of removing facial hair on a woman – so that it never comes back.' Mr M. J., Bristol.

'I propose that an inventor be commissioned to work on a machine that can teach a person of average intelligence the more advanced principles of any topic in, say, a matter of minutes. It would save vast sums of money presently being spent on new schools and universities. The machine should be capable of doing a super-fast "mental transplant" without damaging the recipient's personality.' Mr A. J. R. B., London E.

'Scientists should be working like beavers to discover a way of broadening the human intelligence.' Mr K. G., Sussex.

'Cold air can often bring on attacks of angina pectoris. Could someone invent a portable apparatus through which I could breathe warmed air so that I could go outside in the winter without fear of an angina attack?' Mr M. W. A., Glamorgan.

'Doctors should be working on a pill to control and increase will-power. It is the only cure for those wanting to overcome addictions to drugs, alcohol, and smoking.' L. W., London SW.

Selective slimmer
'My wish is for scientists to perfect a substance that can be injected into unwanted fat to disperse it without dieting. Millions of people are afflicted with more flesh than they want around the waist – but conventional slimming often causes other areas to slim away first. And who wants hollow cheeks at my age?' Mrs H. M., Hertfordshire.

'Since dentists cannot seem to conquer tooth decay, they should find a way to carry out human teeth transplants.' Mrs L. S., Oregon.

'Doctors should devote time and energy to find an effective cure for premature hair loss in men.' D. J. R., Liverpool.

Portable printer

'Please invent a small, portable plaque which, when it receives the spoken word, prints the words for the deaf to see. It should have a small battery that can be switched off when not in use.' G. H. L., Cornwall.

'Some people cannot wear glass eyes and they don't want to go around with frosted glass in their spectacles to hide their disfigurement. Could an optician invent a spectacle lens with an artificial eye fitted behind it? It would have to match the "good" eye – and be unbreakable.' Mrs G. J. H., Hertford.

'My one wish is for medical research workers to find a cure for agoraphobia. If you've never heard of it, it is the fear of going into open spaces. Thousands of people suffer from this ghastly illness. Can't something be done?' (Unsigned.)

'Until scientists find a way to make nerve ends grow they will never conquer the problem of paralysis. If as much money was spent on this line of inquiry as was spent getting to the Moon it would do a lot more good for humanity.' Mr F. N., Los Angeles.

'The greatest need now is a machine to diagnose diseases of the heart.' Mrs K. P. (SRN), Kent.

'Somebody should invent a small electrical attachment that can be worn on the head to treat epileptic attacks the moment they start. It would be able to pick up abnormal brain waves and immediately generate a pulse of electricity to jerk them back to normality again. Instead of a fit, an epileptic would only feel slightly dizzy.' R. S. (aged 13), Suffolk.

Tranquil waves

'What needs to be invented in this frantic materialistic era is a machine that records the pattern of one's brain currents during a state of calmness and relaxation. Then during moments of stress and pressure they could be fed back into the brain. It would be better than drugs.' B. M., Essex.

'If doctors want to reduce the amount of misery and heartache in the world, they will find a cure for eczema – and soon.' Mrs G. B., Essex.

'I would like to order a surgeon to invent an operation to replace all deformed arthritic joints with plastic substitutes.' Mr Z. D., Brussels.
(*Editor's note:* Though *all* joints cannot yet be replaced, surgeons can already implant plastic knees, hips, fingers, elbows and shoulders.)

Coronary indicator

'There must be a way to spot an impending heart attack. Surely it must be possible to make electro-cardiographs that are sensitive enough to show up the clues that must be there. My husband had a perfectly "normal" cardiogram 12 hours before he collapsed with a coronary.' Mrs K. W., Lincolnshire.

Time traveller

'In this day and age scientists should find a way to cure the effects of travelling through time zones. It is easy enough to alter your wrist-watch; it is not so easy to alter your biological clock.' Wing Commander T. S. (Retd), Geneva.

'What we want is a cure for the common cold and 'flu.' Miss L. O., New York City.

'I would be grateful to any doctor who produced a drug which would permanently change "mousy" hair to a definite colour. Constant colouring of the hair with dyes makes it very unhealthy.' Mrs R. U., Rome.

'Jaundice is one of the worst illnesses you can get. Is anyone doing any serious work to find a cure?' Mr F. T., Barnes, London.

'Travel pills don't always work so I would be grateful to any scientist who can find a cure for sea sickness.' Mr D. W. T., Hampshire.

Bad feet
'This may sound silly, but doctors should concentrate on the problems of bunions. They are killing me!' Mrs M. C., Glasgow.

'The sooner doctors and scientists understand the problems of organ rejection, the sooner most of our immediate problems will be over. If you can only stop a kidney or a heart being rejected then not only will lives be saved but other lives will be made bearable, too. I mean by that doctors will have found a cure for hayfever. I am sure all these problems have a single answer.' Mrs W. G., Birmingham.

'Can't anybody find a real cure for dandruff?' Mr P. L. T., Wiltshire.

Finally, and perhaps surprisingly in what many regard as a materialistic age, a suggestion that was echoed many times over. Mrs C. H. from the Netherlands sums up a lot of people's feelings when she says:

'A system should be created to ensure that discoveries and inventions that benefit mankind must be produced at absolutely minimum cost in the shortest possible time so as to do the greatest good for the largest majority. Further, these inventions should not be exploited for profit.'

Picture Credits

Acknowledgement is due to the following for permission to reproduce illustrations on the pages listed:

9–12, N.A.S.A.; 15, 18, 25, 27 and 30, N.A.S.A.; 40–41, B. J. Nixon, Deepsea Ventures Inc.; 42, top left, top right, Scripps Institute of Oceanography, bottom, Ocean Science and Engineering; 44, Rick-Fot; 47, top, U.S. Navy, bottom, Japanese Defence Agency; 51, Cocean, France; 55–56, Cliché Sogréah, Grenoble; 60, Rick-Fot; 62–63, U.S. Navy; 64, top, Tommy Wiberg, Expressen, bottom, E. Dawson Strange; 66, Dick Clarke, International Underwater Explorers Society; 67–68, W.F.A. Marine Fish Cultivation Unit, Ardtoe; 71, top, Experimental Hyperbaric Centre, Marseilles; 76, Thomson-C.S.F. (France); 81, Flower Gardens Ocean Research Centre; 84–85, Shell; 96, University of California, San Diego; 107, Thomson-C.S.F. (France); 112, C. A. V. Acton; 114, top, Krauss-Maffei, middle, E. W. Kendrew; 115, top, British Leyland, bottom, British Railways Board; 116–117, German Post Office; 122–123, National Scientific Research Centre, Montlouis, France; 124, Cambridge Scientific Instruments Ltd.; 126, top, Department of Microbiology, University of Birmingham, bottom, T. E. Thompson, University of Bristol; 127, T. E. Thompson, University of Bristol; 128, top, Dr C. W. Arbuthnott, Department of Physiology, Aberdeen University, bottom, Cambridge Scientific Instruments Ltd.; 129, Dr Stuart Agrell, Department of Minerology, University of Cambridge; 130–131, T. E. Thompson, University of Bristol; 133, Cambridge Scientific Instruments Ltd.; 134, top, Unilever Ltd., Isleworth, bottom, Dr Stuart Agrell, Department of Minerology, University of Cambridge; 135, Cambridge Scientific Instruments Ltd.; 136, Centre for Materials Science, University of Birmingham; 137, top, Douglas Coates, Royal Radar Establishment, Malvern, bottom, I.C.I. Mond Division; 138, top, British Museum (Natural History), bottom, T. E. Thompson, E. Hull, University of Bristol; 139, T. E. Thompson, E. Hull, University of Bristol; 140, I.C.I. Mond Division; 141–144, Pest Infestation Control Laboratory; 145, Department of Botany, University of Birmingham; 146–147, Institute of Tree Biology, Natural Environment Research Council; 154–155, Cambridge Scientific Instruments Ltd.; 158, Hutchins Photography Inc., Belmont, Massachusetts; 162, top, Westinghouse, middle, British Mines Research Establishment, bottom, Rees Group; 164, Rees Group; 168, top, Pilkington Glass Company; 169, Southend Hospital; 170, Monsanto Company; 172, B.O.A.C.; 173, Product Engineering Research Association of Great Britain; 178, E.M.I. Electronics Ltd.; 181–182, Bell Laboratories; 184–186, Magnaflux Ltd.; 187, top, Rolls-Royce, bottom, Magnaflux Ltd.; 188, Keith Morris, *World Medicine*; 190, Slendertone Ltd.; 193–194, Keith Morris, *World Medicine*; 200, Hugh Davies; 201, by permission of Edward Ihnatowicz; 202, Stereo photographs, Margaret Benyon; 204, A.E.I. Scientific Apparatus Ltd.; 207, Camera Press, London; 209, Camera Press, London; 210, Paul Brierley; 215–216, Burden Neurological Institute; 218, Dr R. G. Bickford; 220–222, Dr R. G. Bickford; 223, Basel Institute for Immunology; 224, Paul Brierley; 232, Professor Blechschmidt, *From Egg to Embryo* published by Deutsche Verlagsanstalt, Stuttgart, 1969; 234, top, Dr T. G. Baker, Royal Infirmary, Edinburgh, bottom, Thomas H. Clewe, Delta Primate Center, Covington, Los Angeles; 236,

Dr R. F. Chen, National Institute of Health, Bethesda, Maryland; 237, top, National Heart and Lung Institute, Bethesda, Maryland, bottom, Ralph Bredland, National Institute of Health, Bethesda; 238, top, National Institute of Health, Bethesda, bottom, Ralph Bredland, National Institute of Health, Bethesda; 240, National Institute of Health, Bethesda; 241, Vickers Ltd.; 243, British Diabetic Association; 245, M.R.C. Cyclotron Unit, Hammersmith Hospital; 247, inset, J. Arthur Rank; 248, Barif and Lambard; 250, N.A.S.A. Ames Research Center, Moffett Field, California; 253, P. Larsen, World Health Organisation; 258, bottom, Royal Aircraft Establishment, Farnborough.

Drawings on pages 260, 262–265, 268, 270 are by Larry.

Late picture credit *Tomorrow's World*, Volume I (published 1970): page 229, Sculpture by Anne Stern, colour photographs by Barry Stern.